WHAT'S SEX GOT TO DO WITH IT?

WHAT'S SEX GOT TO DO WITH IT?

Darwin, Love, Lust, and the Anthropocene

Heather Remoff

First published in 2022 by
Shepheard-Walwyn (Publishers) Ltd
107 Parkway House, Sheen Lane,
London SW14 8LS
www.shepheardwalwyn.com
www.ethicaleconomics.org.uk

British Library Cataloguing in Publication Data
A catalogue record of this book is available from the British Library

ISBN: 978-085683-5483

Typeset by RefineCatch Limited, Bungay, Suffolk

For Ingrid Gallagher and Craig Fowler
and
their families

Contents

Preface

Challenging conventional wisdom is never easy. Arguing for a need to shift the Darwinian paradigm is a bit like walking into a roomful of esteemed physicists and announcing that there is no such thing as gravity. Nonetheless, a handful of evolutionary biologists quietly have been unveiling discoveries that cry out for a reevaluation of some of Darwin's central tenets. In my opening chapters, I begin cautiously by describing recent experiments with *non-human* species that refute longstanding assumptions about the pace of evolutionary adaptations. However, this is a book about the evolution of *Homo sapiens*, and a sense of urgency about what we're doing to the planet moves me quickly past research with lizards and snail kites and into a discussion of my own species.

My initial interest in the works of Charles Darwin was inspired by curiosity about the role of female choice in shaping the human animal. Although my Ph.D. is in anthropology, evolutionary theory quickly became my passion. Forty years of thinking, reading and researching the subject force me to conclude that the failure of modern biologists to reconsider Darwin's work in light of female behavior and recent genetic discoveries has left us with a deeply flawed theory of human evolution. *What's Sex Got To Do with It? Darwin, Love, Lust, and the Anthropocene* employs a female perspective to set the record straight on Darwin's theory of sexual selection, the somewhat shaky foundation on which our current assumptions about human nature are based.

Because we're now in the midst of two existential crises—climate change and economic inequality—caused by human behavior, we have no time to waste. If we are to be successful in addressing these challenges, our scientific postulates must be based on empirical evidence. The smoldering skies over San Francisco convince me that if we're going to survive, we need to incorporate recent genetic breakthroughs into our understanding of what makes *Homo sapiens* tick. It's time to come to terms with the double-edged sword at the heart of our exceptionalism. In developing an updated theory of human evolution, I've drawn on insights from a number of disciplines—biology, economics, and anthropology. My initial understanding was inspired by the hundreds of hours I spent interviewing women about factors influencing their choice of sexual partners. My doctoral research convinced me that neither Darwin nor I knew nearly enough about economics. Correcting for that deficit took me on a journey that exposed the evolutionary importance of differences between the way humans control access to resources and the way that all other species on the planet do. Below is a brief summary of the central topics this book covers.

The power inherent in sexual selection and female choice

Ever since Darwin published his seminal work, biologists have agreed that sexual selection is driven by female choice. The failure to pay close attention to the strategies and traits influencing the choices a female makes has been consistent across species but has nowhere resulted in as much evolutionary misunderstanding as it has in the human case. Since *What's Sex Got to Do with It* is first and foremost about human evolution, the book brings evolutionary theory to life by periodically turning to examples that remind the reader that the traits defining us as *Homo sapiens* continue to contribute to a man's sex appeal.

Chromosome fusion

It is so difficult for biologists to observe data that fall outside Darwin's original beam of light that some rather dramatic, theory-altering recent genetic breakthroughs have not received as much attention as they should. Why hasn't the discovery of an astonishing evolutionary mutation—the end-to-end fusion that combined great ape chromosomes 2A and 2B to create human chromosome 2—revolutionized biology departments around the world? Almost overnight, we went from being a species with 24 chromosome pairs to one with 23. This is the evolutionary sleight of hand that established *Homo sapiens* as a distinct species, forever separated—both behaviorally and biologically—from our non-human primate cousins. Our speciation event is headline-making stuff. Why hasn't it made headlines? Because when it comes to human evolution, Darwinian theory is the third rail that biologists hesitate to touch. The genetic correlates most likely to undergo alterations in the way in which they manifest are those occurring in the region of the fusion. Interestingly enough, the traits that were most significantly impacted by our speciation event were those relating to sex and the brain.

Concealed ovulation and continuous sexual receptivity

I hypothesize that the mutation resulting in concealed ovulation and continuous sexual receptivity was part and parcel of the initial end-to-end fusion, which combined great ape chromosomes 2A and 2B to create human chromosome 2. This unique-to-humans reproductive innovation was not the result of slow and steady change driven by natural selection but was born at the moment of our original speciation event and gave our ancestral Eve the power to direct the evolutionary sprint from great ape to modern human. Adaptations driven by sexual selection are characterized by a velocity not possible under a system of natural selection as described by Darwin. The differential rates of speed available to natural selection versus sexual selection are a recurring theme throughout the book.

Symbolic control of resources

Another unique-to-humans innovation—a shift in the way the human brain processes and encodes information—is likely to have accompanied our birth as a primate with 23 chromosome pairs. I propose that a possibly tiny neurological alteration facilitated *symbolic* representations and set the stage for the evolution of language, arguably the most important characteristic defining humanity. Human intelligence and language ability enabled early humans to take an innovative approach to controlling the resources necessary to their survival and that of their offspring. The members of all species *except humans* use their own bodies to stake out turf containing the food and shelter they need to live. Humans control resources *symbolically*. This distortion of the notion of property rights makes it possible for individual members of our species to own far more of the world's wealth than they can possibly use in their lifetimes or those of countless generations of their descendants. A trait that was once adaptive has morphed into one responsible for environmental destruction and the worldwide growth in inequality. Human economic behavior got us into this fix, and it can get us out, but only if our search for answers is grounded in an accurate understanding of our evolved species-specific traits. In the book's final chapters, I employ our new understanding of human nature to suggest efficient policies that would reverse climate damage and promote economic justice.

Confirmation Bias

In *The Sixth Extinction,* Elizabeth Kolbert makes a compelling case for an extinction event of massive proportions being humankind's most lasting legacy unless, as it states on the back cover, we are able to rethink the fundamental question of what it means to be human. My hope is that by the time readers have turned the last page in *What's Sex Got To Do with It,* they will have come to a better understanding of who we are and how we evolved. We are a young species, one that hasn't stood the test of time, and unless and until we are able to alter some of our fairly recent behavior, it is a test we appear doomed to fail. But our reign on Earth doesn't have to end this way. It's possible to design a system of policy rewards that shifts the focus of the very human traits that got us into our current predicament and puts them to work getting us out. It all begins with an understanding of what those traits are.

Once again, Kolbert alerts us to a barrier standing between humans and a future in any way resembling the world we now enjoy. Confirmation bias, the inability to see something that confounds our expectations, explains the difficulty early researchers had convincing peers of the importance of the break in the fossil record "between the plants and animals found in rocks from the late Cretaceous period and those found directly above . . .". Darwin himself was one of the first to disavow the validity of the evidence those fossils offered. It was not until the gradual acceptance of the Alvarez hypothesis in the 1980s that the significance of the Cretaceous extinction became part of our common understanding of the evolution of life on this planet. As Kolbert points out, it takes a paradigm shift in order for people to process disruptive information. I'm fascinated that the theory of natural selection was already so firmly ingrained in the public imagination by 1859 that neither researchers of Darwin's era nor those of ours have—until relatively recently—been able to *see* evidence of a massive extinction event, even though it was right there, physically in front of their eyes, clearly inscribed in the geological fossil record. Since the prevailing wisdom was, and generally still is, that evolution proceeds slowly, it was almost impossible for them to concede that the flip side, the extinction part of the equation, could be devastatingly swift.

It's equally difficult for modern biologists to allow that *speciation* events might happen as suddenly as those characterizing extinction. I can think of no other explanation for the general failure of the scientific community to celebrate and widely publicize Jorge Yunis' and Om Prakash's 1982 discovery of the chromosome fusion that appears to be the true origin of our species. If I accomplish nothing else, I want their insights to inspire discussion and become an established part of the conversation. What do I expect from the reader I imagine while crafting these words? Only that you keep an open mind. Discover the thrill of exploring ancient questions from a slightly different angle. Have faith in the value of your own ideas and explanations. Have fun thinking about these things. It's a quintessentially human activity. From our earliest beginnings, we've wanted to know who we were and how we came to be. Let's put our heads together and see if we can come up with answers that allow us to reaffirm our belief in ourselves as the *clever* hominid.

Chapter 1: In the Beginning

What's sex got to do with it? Pretty much everything—at least if you're a member of a sexually reproducing species, and humans most definitely fall into that category. Simply stated, sexual reproduction provides a store of genes that can be traded and mixed about whenever species blessed with individuals of both the male and female variety need to be quick on their feet in order to survive environmental changes. While all sexually reproducing species have taken advantage of their expanded genetic opportunity and come up with a number of pretty amazing adaptations, humans alone seem to have figured out a further variation on this innovative reproductive game, one that has somehow moved us into unexplored territory. Other animals use sex to engineer a modest bit of genetic reshuffling, but those of us who are *Homo sapiens* have managed to pull off something quite spectacular. Just like many species on the planet, our adventure started with sex, but, unlike the others, we ended up with big brains, a way with words, and transformative technological skills. What were ancestral humans doing sexually that those other species weren't? How did we manage to jumpstart our own evolution and turn the Darwinian process on its head? How is it that we figured out how to shape the world to our own ends, rather than waiting around for natural selection to adapt us to it? We were able to pull all this off because *Homo sapiens* has a unique approach to sex that lies at the heart of our sprint to the head of the class, at least when judged by those who view evolution as a purely progressive enterprise. But before exploring the highly unusual mutation that allowed human females to turn sexual selection into a superpower—think of it as evolution on steroids, on estrogen and progesterone—we would do well to be clear about just how sexual selection differs from natural selection. Not all biologists agree that it does. Many dismiss sexual selection as simply a subset of natural selection and feel that the theory of natural selection is sufficient to explain the origin of species and pretty much everything else about the adaptive process. Why was there even a need to develop a theory of sexual selection? For the answer to that question, we need to turn to Charles Darwin.

Darwin's struggle to explain the beauty of the peacock's tail inspired his embrace of sexual selection. Those who challenged the theory of evolution through natural selection saw the hand of God in traits that appeared to have no demonstrated function other than the ability to delight the eye. If evolutionary theory proposed to remain scientific and not vulnerable to arguments that relied on theological explanations, Darwin had to find a way to account for traits that failed to demonstrate the biological fitness that evolution uses to sort winners from losers. Himself a pigeon breeder, Darwin had firsthand experience with *artificial* selection's ability to highlight features that served no purpose other than to dazzle fellow hobbyists with dramatic displays of form and color not found in the wild rock doves that were the foundation stock. The characteristics the fanciers fancied were purely decorative and would not have served the domestic birds well had they been forced to go head to head against native pigeons in the red-in-tooth-and-claw arena of natural selection. *Artificial* selection could account for beauty without function in the domestic breeds so familiar to Darwin's generation of backyard animal enthusiasts, but that was at the hands of *men* who had both an eye for the exquisite and the agency to bring their visions to fruition. How was Darwin to explain the extremes of beauty sported by male birds in the natural world? In answering this question, it's important to remember that—at least from Darwin's point of view—a major problem with the peacock's tail was that it belonged to the pea*cock* and not the pea*hen*. Prevailing theories of mate choice emphasized male to male competition, with access to the female or females in question going to the male who'd proved triumphant in battle. Darwin, who was very much a man of his time when it came to his assessment of male superiority, had a dim view of the intelligence of females in all species, even his own. It's an understatement to say that it was not easy for him to shift his focus from male strength to female choice, but the splendor of that tail eventually forced him to give the peahen her due. Not only that, but he also managed to move beyond birds and postulate that early humans, even those of the female variety, had once possessed sufficient critical judgment to influence the evolution of the various races of man, hence the title of his eventual book on the subject: *The Descent of Man and Selection in Relation to Sex*.

I first read *The Origin of Species by Means of Natural Selection* and *The Descent of Man* in the 1970s, but not until after the emergent field of sociobiology had captivated me with evolutionary explanations grounded in an understanding of genetics that would not have been part of the research material available to Darwin. (b. February 12, 1809–d. April 19, 1882) As a result, by the time I began turning the pages of his book on sexual selection, I read far more into it than the great man may have himself intended. I didn't need Darwin to convince me that sexual selection was powered by female choice. Robert Trivers, the biologist whose research established the foundation on which subsequent sociobiological analyses would be based and the outside member of my dissertation committee, already had. Certain that women's sexual decisions were more complex than those involving nothing other than simple assessments of physical appearance, part of my doctoral research in anthropology consisted of tape-recording hundreds of hours of interviews with women who described the traits that had influenced their choice of specific romantic, sexual, and reproductive partners. I wanted to know exactly what characteristics triggered a woman's interest in the men she found attractive. Although I was investigating sexual selection, I accepted natural selection's premise that the most successful members of a species would produce greater numbers of offspring than their less fortunate rivals and hypothesized that the women in my study would have more children with the men they described as successful than they would with those they viewed less favorably. I also predicted that women would be more conscientious users of birth control with men they considered to be unacceptable reproductive partners. My data not only failed to support either hypothesis, it actually hinted that quite the opposite might be the case. I'll return to some of the implications of that research in subsequent chapters. The point I want to make here concerns the reaction to my work when I presented my findings at conferences in related fields of science.

Because I was talking about women's sexual behavior at gatherings where I was likely to be scheduled against presentations analyzing the mating habits of drosophila and other species that didn't require clearance by the Committee to Protect Human Subjects, I usually spoke in auditoriums packed with predominantly male scientists. They were especially eager to learn what traits in men aroused sexual desire in women. I had no difficulty holding the attention of my audiences. In the question and answer sessions that followed, it was inevitable that when the microphone was handed to the first guy who'd raised his hand, he would not ask a question but would instead make a statement, "This is all very interesting, but it's not sexual selection."

This happened in presentation after presentation. Why was I never prepared for the bit of instruction that, after encountering it a few times, I should have come to expect? The comment puzzled me enough that I quickly moved past it and never managed to respond in a way that might at least have helped clarify the origin of the confusion between us. I was simply bewildered by the failure of my fellow researchers to see something that was patently obvious to me. I couldn't understand why my documentation of the characteristics influencing a woman's choice of reproductive partners somehow fell outside their understanding of how a system of sexual selection worked. But, in fact, I was the one who had missed something obvious. As a relative newcomer to the field of evolutionary biology, I'd never internalized a cardinal bit of conventional wisdom regarding the limits of this unique evolutionary force. In my enthusiasm for what I thought was Darwin's take on female choice and sexual selection, I had glossed over a central tenet of his thesis: the belief that sexual selection was responsible primarily for secondary sex characteristics, those traits that mark sexual dimorphism and are expressed in either one biological sex or the other. As a woman with two children, I knew that the sexual choices my heterosexual female peers and I made were based on a whole range of factors other than appearance and were related to sexual dimorphism only at a level so basic as to need no articulation. In other words, we could tell the difference between men and women and—at least as far as our *reproductive* behavior went—preferred one over the other.

I wish I'd known then what I know now about the rather widespread misunderstanding among biologists regarding the difference between sexual and natural selection. Perhaps I would have been able to convince the men in the audience that my research actually had something interesting to say about the process of human evolution and was concerned with a bit more than simply providing them with pick-up lines that they later that evening could try out in the local bars. How should I have responded all those many years ago? By pointing out that not even Darwin had been able to fully grasp the revolutionary amount of power sexual selection gives species to shape their own evolutionary destinies. His mistake, and that of many respected biologists working in the field even today, was based on the flawed assumption that sexual selection works *only* on secondary sex characteristics such as the oft-mentioned peacock's tail. I prefer to describe the difference between natural selection and sexual selection as this: sexual selection determines *who* gets born, natural selection determines *what* happens to them once they embark on the struggle to survive.

This definition is not so very different from one Darwin gave in *The Descent of Man and Selection in Relation to Sex*. "Sexual selection depends on the success of certain individuals over others of the same sex, in relation to the propagation of the species; while natural selection depends on the success of both sexes, at all ages, in relation to the general conditions of life." Even though explaining the existence of that infamous peacock's tail is what inspired Darwin to develop his theory of sexual selection, there really is no need to make sexual dimorphism an essential part of the distinction between sexual and natural selection. That said, I should be grateful to the troublesome tail. It was its existence in males and not in females that forced Darwin's hand when it came to acknowledging the role of female choice in his new theory of evolution. Given the difficulty Darwin had in crediting females with intellectual agency of any kind, I'm glad that he and the peahen at least shared an aesthetic sense. Since her perception of beauty matched his, Darwin was able to grant, however grudgingly, that she might not be *entirely* bird-brained.

Sexual Selection, the evolutionary theory that is uniquely Darwin's own, was inspired by frustration over his inability to explain the existence of something—the splendor of the peacock's tail—that appeared to him to have *no function*. After all, the theory of natural selection was based on survival value, the advantage that slight modifications in form gave the individuals possessing them over others of their kind. How then was one to explain beauty, something that could not be demonstrated to enhance one's survival, something that appeared to be without a competitive advantage? Darwin and his peers tried. They, and even some modern biologists, have suggested that the ability to unfold a tail large enough to burden the one displaying it might, in and of itself, provide a demonstration of the physical strength and underlying fitness of the male vying for the female's attention. Questions of what constitutes beauty aside, there is a fairly obvious—at least to me—fallacy at the heart of Darwin's theory of sexual selection: the assumption that the admittedly gorgeous peacock's tail has no function. *It has exactly the same function that all sexually dimorphic traits, beautiful or not, share; it enables the females making reproductive choices to tell male from female.*

From that initial error, Darwin, and evolutionary biologists ever since, have slipped into a second fallacy, one that does not even logically follow from the first. What was Darwin's second mistake? His belief that sexual selection has the ability to shape *only* those features relating to and enhancing markers of sexual dimorphism, or what are commonly called secondary sex characteristics. I guess the confusion is understandable. After all, reproductively successful peahens always mate with birds sporting that magnificent fan of feathers. I would even agree that they can and do modify the expression of this sexually dimorphic trait by consistently choosing males with the largest display. I don't doubt that the cocks become more pleasing to the hens when the impact of the spectacle is further enhanced by intricate and colorful patterns. However, the bottom line is that peahens are mostly engaged in picking the males that are best at advertising their *maleness*. A peacock's big beautiful tail constitutes a super-stimulus. Later in this book, I'll discuss the appeal of the super-stimulus. However, when we're talking about the traits that trigger courtship behavior, I'm convinced that the *initial* thing peahens are selecting for is maleness. And since bird genitalia are *internal*, birds need *external* markers of sex in a way that species with external genitalia do not. Primates, for example, aren't as reliant on secondary sex traits in order to figure out who's got the sperm and who's got the eggs; their primary sex characteristics are on the outside. However, peahens are forced to rely on secondary sex characteristics in order to tell the boys from the girls. A peahen's choice *correlates* to the trait identifying the sex of the bird. It doesn't *cause* the biological sex with which the trait is associated. That's not to say that the feature identifying maleness might not have started with one tiny feather and been sculpted by female choice into something that satisfied even Darwin's sense of beauty. However, the females in question required something in addition to pretty feathers. They needed to lay fertile eggs, and only sex with pea*cocks* could provide that outcome. Peahens who are sexually turned on by the more subtle coloration of their sister pea*hens* are not going to get their genes into the gene pool of the next generation. That's pretty much the heart of the matter. Same-sex sexual behavior may be perfectly desirable in terms of the happiness of the particular birds in question and the social relationships among birds within the flock; it simply doesn't give *evolution* much to work with when practiced by members of a sexually reproducing species.

By limiting our understanding of how sexual selection works and overlooking the full range of traits on which female choice might potentially operate, we seriously limit our understanding of sexual selection's power to enable sexually reproducing species to rapidly adapt. We underestimate sexual selection's role in driving evolution. Once there are two sexes, once we have a sexually reproducing species, females are able to select for *any* trait that renders the male in question more fit and appealing than his competition; more intelligent; better able to exploit resources; stronger with a demonstrated ability to survive; more willing to share food and help with rearing the kids; more accomplished at establishing territorial displays; better able to sing the most melodious song; better able to build her shelter from the storm; perhaps prettier, better able to please her eye; better able to employ words to promise commitment into the future; better able to tailor his sexual passion so that it feels specific to her; and yes, better able than anyone else to make her laugh. The expression of some of these traits might be dimorphic along sexual lines. However, it's not simply about secondary sex characteristics. The main purpose those serve is to enable participants in the mating game to tell the lads from the lasses. Once females get past the business of choosing a potential mate that belongs to the correct species, they move on to selecting one that is the reproductively appropriate sex, and only then can they get down to the business of selecting the traits that truly enhance that male's fitness and that will end up enhancing the fitness of any offspring resulting from the union.

It may be true that—at least when compared to humans—the females of other species aren't making choices that have been demonstrated to impact traits beyond those associated with male versus female: appearance, song, mating calls of all sorts, pheromones, etc. (Of course, when the *assumption* is that female choice results *only* in sexual dimorphic traits, there isn't much incentive to undertake research correlating mate selection to anything that doesn't fall within that rather narrow rubric.) Given the brouhaha following the publication of *The Origin of Species*, it's understandable that Darwin saved his discussion of human evolution for *The Descent of Man*. While the delay may have been, at least in part, inspired by the demonstrated reluctance of the general public and even many scientists to embrace an evolutionary theory that appeared to rob humans of their divine origins, the discussion of human evolution in the same volume that also unveiled Darwin's theory of sexual selection is appropriate in ways he could not possibly have imagined. Darwin may have failed to understand that sexual selection could impact traits other than those relating to appearance, but he at least introduced us to the concept of female choice. And since it was only in the early 1980s that scientists became aware of the fusion that combined great ape chromosomes 2A and 2B into human chromosome 2 (giving humans 23 chromosome pairs and great apes 24), there's no way Darwin could be faulted for not considering this rather abrupt mutation as a potential speciation event. Focused on the gradual nature of the evolutionary process, one that required many millions of years, Darwin wasn't looking for the dramatic introduction of new species. He established the deep nature of evolutionary time. Most scientists agree that fully developed humans did not suddenly appear in some Garden of Eden. That said, we are a fairly recent innovation in a primate line going back at least 55 million years. A lot has happened to *Homo sapiens* in a mere 300,000 years. Darwin didn't have radiocarbon dating at his disposal, but he knew that the difference between humans and even our closest primate relatives was significant enough to warrant a separate treatise. Since I strongly suspect that the fusion that established a chromosomal boundary between us and the great apes also resulted in a change in the *reproductive biology* of human females, I delight in the serendipity that made sexual selection and female choice part of Darwin's discussion of human evolution.

There is a bio-behavioral difference that separates humans not only from the great apes but also from other mammals. And it is this unique trait that has enabled women to raise sexual selection to a high art. While Darwin and evolutionary theorists from his grandfather, Erasmus Darwin, on down were busy championing the importance of male-to-male competition in mate selection and struggling to figure out how to explain the beauty of male birds without having to give females an undue amount of credit, there was one philosopher and political theorist who understood why human mate choice wasn't characterized by men fighting with each other over access to women. In taking issue with Thomas Hobbs' description of the battle between male animals over possession of the females as being indicative of the brutal nature of interactions between primitive humans, Jean Jacques Rousseau identified the reproductive distinction that has given women so much power in the courtship arena. "Rousseau repudiated Hobbs' animal analogy by stressing the significant difference in the reproductive physiology of human females.

Women, unlike female animals, do not have alternating periods of 'Passion and indifference' and thus are always sexually available. 'We cannot, therefore, conclude from the Battles of certain animals for the possession of their Females, that the same would be the Case of Man in a state of Nature'" (Eveleen Richards, 2017, quoting Rousseau). Although I tend to question Rousseau's belief in humanity's peaceful origins, his focus on the significance of the lack of a visible estrus cycle in ancestral humans is one that I share. Rousseau was correct in sensing the way in which concealed ovulation and continuous sexual receptivity would reduce the importance of physical battles between men over access to women. Concealed ovulation endows human females with a heightened role in matters of courtship. It's not that men have *no* power; it's simply that women are able to exercise their options with enough finesse and subtlety that they control the pace and complexity of the mating dance. If a man lacks interest in the woman who has identified him as a potential sexual partner, his failure to respond effectively negates her choice. At the risk of engaging in dangerous stereotypes, men evaluate women on fewer traits than women employ when determining a man's desirability as a partner. Appearance and sexual availability can pretty quickly trigger desire in men. Women tend to need a bit more wooing, a bit more sweet-talking. The failure to examine all the traits they evaluate before choosing one man over another has robbed us of some interesting insights into the nature of our evolutionary history. It is, after all, primarily our skill with language that sets *Homo sapiens* apart from all the other species on the planet, and tracing its origins remains a challenge for both linguists and biologists. The common misconception that sexual selection results only in traits that are sexually dimorphic has deterred linguists from even considering the role of female choice in driving the evolution of language and has thrown an unnecessary roadblock between us and our understanding of the development of all things human.

A few years ago, infected by a colleague's expressed interest in the evolution of this specifically human claim to fame, I worked my way through a number of academic texts and popular books on language that had been published in the years since I'd written my dissertation and had first speculated on its importance as a factor influencing female choice. Every one of those books gave me some new angle to consider, but it was Terrence Deacon's *The Symbolic Species: The Co-evolution of Language and the Brain* that I quickly dubbed the best of the bunch. There is no theorist on the subject of language I respect more. Therefore, it's weirdly appropriate that only six months ago, while reading a scholarly article written by him in 2010 that it, *at last,* dawned on me how deeply into our Darwinian understanding that flawed interpretation regarding sexual selection had wormed its way. In discussing the possible role this unique force might have played driving in the evolution of language, Deacon argued that there are "strong reasons for doubting its relevance to this most distinctive of human traits. This is because sexual selection inevitably produces complementary divergence of male and female traits, as is exemplified by peacock tails . . ." *What?* Surely I must have misunderstood the point he was making. I had to reread the passage several times before finally grasping that not even someone as brilliant as Terrence Deacon was immune to being tripped up by a too-narrow focus on secondary sex characteristics. Deacon continued his explanation, "Therefore, accounting for the extravagant complexity of language in terms of sexual selection requires explaining why it lacks this otherwise-ubiquitous mark of extreme sexual dimorphism." It took my high regard for Terrence Deacon and that moment to *finally* force me to confront what had been at the heart of the misunderstanding between me and all those men who had raised their hands to tell me that my research on the factors influencing female mate choice had nothing at all to do with sexual selection. As far as they were concerned, if the traits I was describing weren't sexually dimorphic, they couldn't possibly have been influenced by sexual selection. It would appear that Darwin is not alone in having been hypnotized by staring for far too long into the eyespots adorning those famously magnificent feathers. I don't know how else to explain the almost comic blindness of male research biologists to something that is going on all around them, all the time.

Chapter 2: Why Now?

Over 7,500 books have been written about Charles Darwin. Other than a bit of score-settling with all those "mansplainers" from the late 1970s, what makes me think the world needs another one? For starters, impressed as I am by the quality of recent books on sexual selection, I think they remain caught in the beauty trap. Darwin also was led astray by an emphasis on physical appearance, by that alluring lure that so often first captures a man's attention. Despite my quarrel with him in this area—male researchers tend to project the factor influencing their own choices onto the females doing the selecting—I stand firmly with him on a more neglected aspect of his theory of human evolution. Sexual selection is the force that has made us who we are. It's no accident that the title of the book in which he explored the evolution of his own species contains the phrase "and Selection in Relation to Sex." Near the end of the preface to the second edition, Darwin briefly expressed puzzlement as to why his theory of sexual selection never seized the public imagination in the same way the theory of natural selection had. He concluded that once naturalists had had time to become familiar with the idea, it would become much more widely accepted. Unfortunately, that has not happened. It's not that modern biologists don't acknowledge the existence of sexual selection; it's that they underestimate its power. Reproductive choices are about so much more than selection based on physical appearance. The human case is especially interesting. Our earliest female ancestors put a new spin on an old game. Reproductively, they were doing something very different from what even our closest primate cousins were doing, and the options made possible by an evolutionary twist that combined continuous sexual receptivity with concealed ovulation can help to explain much of what is unique about human culture, biology, and behavior.

Why does that matter now? Because the ways in which we *Homo sapiens* employ our species-specific behavioral patterns are wreaking a bit of havoc, on both a personal level and on a global scale. I don't know many women—or men—who are particularly happy with the current state of either their sexual or romantic lives. And it's not only at the level of courtship that we hear a whisper telling us that all is not as it could be. Consider our sense of economic security or, more accurately, *insecurity*; it is perhaps only individuals in the wealthiest top 10 percent who express satisfaction with their status. Zoom out.

Move from the personal to the global and consider the worldwide increase in inequality. Now raise the curtain on the Anthropocene. Although the term has not yet been officially recognized, it is widely understood to refer to the current geological age, the one in which human activity has been the dominant influence on climate and the environment. Our impact is generally seen as a negative one, at least if we are evaluating the health of a planet on which our own survival depends. Is everything really going as swimmingly as we'd like to pretend? Perhaps a shift in the lens through which we view ourselves and the way in which we were shaped by the evolutionary process is in order.

The failure to accurately distinguish between natural and sexual selection continues to catch biologists by surprise when their field research documents evolutionary changes that don't appear to be playing by the rules, according to Darwin. A current example made it into the pages of the *New York Times*. A bit of conventional wisdom shared by both scientists and members of the general public is that natural selection *slowly* shapes species over time. One of Darwin's claims to fame is that he established that the geological record offered proof that earth was much older than the biblical accounts of creation claimed. Within this expanded timeframe, species could slowly and steadily evolve into the teeming multiplicity of life forms that characterize our world. Yes, species evolved. All they needed in order to do so was to be granted more than 10,000 years. What then are we to make of the rapid changes in beak and talon size recently documented in a population of North American Snail Kites, a species of hawk found in Florida?

As the article in the *NYT* reports, their numbers were dwindling so dramatically that when the introduction of an invasive species of large snails began to replace the smaller snails on which this endangered species of bird fed, extinction became a clear possibility. As their name would imply, snail kites have specialized curved beaks and long claws that are well suited to harvesting the small apple snails that share an ecosystem with them in the Florida Everglades. Because the invading snails were much larger and more difficult for the relatively small-billed kites to consume, ornithologists predicted that the very success of the larger species of snail would spell doom, not only to their smaller snail cousins but also to the birds that relied on them for a source of food.

But something unexpected happened on the way to the demise of this population of snail kites: the birds almost immediately evolved beaks that enabled them to take full advantage of the menu provided by their new snail neighbors. What do we mean by *almost immediately*? Are we cutting the time needed for significant change down from a couple of thousand years of sculpting at the hand of natural selection to a mere hundred years or so? Nope. The rate of speed is even more impressive than that. In only eleven years of collecting morphological data, the researchers documented an average growth in beak and body size of eight percent. Some birds grew by as much as twelve percent in both measurements. Considering that snail kites live about eight years, this increase happened in approximately two generations. Somehow, newly hatched snail kites whose parents had been able to out-compete other birds in the face of the increasingly scarce resource of native snails managed to pull off Darwin's vision as influenced by Thomas Malthus. Was this an example of nature red in tooth and claw—or, in this case, beak and talon—driving evolutionary change? Hardly. The endangered snail kite population was no longer declining but had actually *increased* threefold in the thirteen years since the larger snails had invaded, and it all appears to have been accomplished fairly peacefully. The food supplies were no longer scarce. The birds weren't fighting to the death in winner-take-all fashion. The larger birds had triumphed, not by killing their fellows, but by somehow managing to change *themselves*. Not only that, but the changes were more than simply morphological, the result of being well-fed, and appeared to correlate to genetic changes. Large-beaked birds laid eggs that hatched out large-beaked offspring. Ecologist Robert Fletcher, Jr., an author of the study, admits to being very surprised at this demonstrated ability of large-bodied animals to rapidly keep up with and exploit changes in their ecosystem.

In Michael J. Ryan's *A Taste for the Beautiful: The Evolution of Attraction*, one of the wonderful recent books on sexual selection, he cites several experiments that document an almost-built-in female preference for "more" that seems to exist across many species, my favorite being the aptly named widowbird found in some regions of Africa. Ryan saw his first widowbird while in Kenya, studying the heart-nosed bat. His initial impression of a male widowbird skimming above a tall field of grass was that he was witnessing a small black bird fleeing a much-larger predator. Only when his attention was fully engaged did Ryan realize that what he was actually seeing was the flight of an otherwise nondescript bird whose tail was so much longer than its wingspan that any pattern of flight or altitude other than what the bird was executing would have been impossible. In other words, the very excess of a trait that researchers had documented as being highly attractive to females made it vulnerable to predation and likely to make "widows" of the females who, lacking long tails themselves, nonetheless found this fabulous folly of feathers irresistible in their reproductive partners. How is one to explain the evolution of such a trait? It's the question that Darwin found so confounding when he considered the peacock's tail. Theorists have speculated that the very strength required to support something such as a peacock's tail or the outsized antlers of the now-extinct Irish Elk demonstrate the underlying of the health and vigor of the animal displaying them. I'm less convinced by those arguments than I am by

Ryan's emphasis on the force of the super-stimulus *more*. I certainly documented the ways in which extravagant displays dazzled the women I interviewed. Unfortunately, the excessive consumption that characterizes human behavior may prove to be as maladaptive to us as eighty-eight pound antlers were to the Irish Elk. Holding the potentially erotic allure of demonstrations of "more" in mind, let's speculate on how that might have played a role in the rapid evolution of snail kites.

Whenever I come across research that documents a species evolving much more rapidly than the scientists doing the studies would have anticipated, I suspect that natural selection and sexual selection are working in tandem. In the case of the snail kites, there are probably a number of pre-existing genes that influence beak size. I don't think these raptors would have had to wait for a suddenly adaptive mutation that enabled them to feast on the outsized invaders. The beauty of sexual reproduction is that it serves as a store of genes, not all of which are being phenotypically expressed in any given individual at any given moment in time. However, when conditions change, such as when a larger invasive species of snail suddenly moves into the territory, any slight variation in beak size that renders the bird expressing it better able to dine on the intruders would not only enable the bird with the large beak to stay alive but there's also a good chance his appearance and behavior would capture the attention of females making the reproductive choices.

Snail kites are a species of communal hawk in which all adult members harvest snails from shallow freshwater. Skimming low over the surface, they use their talons to lift their prey and carry it to an exposed branch, where they pull the mollusk from the shell with their beaks. Fishing ability and skill at extracting the snail from its shell are readily visible to the other birds in these nesting colonies. Both sexes hunt and are involved in feeding and rearing the young. Natural selection will favor males *and* females that have beaks and talons of sufficient size to exploit a novel food source. Once the snail is lifted from the water and the hawks have plucked out the meat, they can either eat it themselves or fly off to stuff it into the waiting beaks of their offspring back in the nest. There's a potential selection pressure for size even within the nest. Birds typically feed the baby whose mouth is opened the widest. In normal situations, the fullness of the stomach directly influences how far a nestling can spread its jaws, guaranteeing a fairly equitable distribution of calories among the youngsters. It's easy to speculate on the competitive advantage of a young snail kite whose slightly larger beak makes it appear to be perpetually hungry. After harvesting a snail, a male hawk has an option not employed by the females; he can choose to present it to a prospective mate in a display of courtship feeding.

The above descriptions demonstrate the interplay between natural and sexual selection pressures. Courtship feeding is, of course, a reproductive strategy that enhances a male's chances of executing a successful copulation. This is sexual selection pure and simple, and in the course of coding the taped transcripts of the women I interviewed, I documented its existence even in humans. It was a rare woman who had sex with a man who had not fed her first. In Ryan's book, *A Taste for the Beautiful*, he ties a female guppy's preference for the color orange in their prospective mates to sensory exploitation in which the brain is primed in ways that quickly get co-opted for sex. Guppies love to dine on orange fruit. No surprise, I guess, that females find orange such an attractive feature in the coloration of the males they favor.

I don't doubt that females select for beauty, but argue that we should be willing to consider that they also are capable of selecting for traits that are much more directly related to survival skills. If a female is clever enough to notice something as frivolous as tail length, she is certainly clever enough to notice when a male of her species demonstrates skill at exploiting a novel resource, such as the sudden invasion of large snails. This is particularly true when snails are the primary food in a species where both parents assume responsibility for providing for the young. Is a large beak more aesthetically appealing than a small one? Perhaps, at least if Ryan's premise about the seductive property of *more* proves to be true. He documents its existence in a range of species. His examples of experiments with widowbirds, swordtails, platyfish, grass finches, and túngara frogs convince me that *something* is going on. Is enhanced size all that a beak needs in order to be considered beautiful? Perhaps. Perhaps not. But I don't think anyone will challenge the notion that there is also something seriously beautiful about a full stomach and the promise of being able to raise your offspring to adulthood.

Courtship feeding in humans was an accidental discovery in my own research. I certainly wasn't coding for it. But—once identified—it was, of course, everywhere. One might argue that going out to dinner is such an ingrained part of the dating culture in the United States that I should not have been surprised by the finding. And, in fact, I wasn't *surprised* when a woman began reciting every detail of a meal she'd shared with a lover, but I was slightly *bored*. I was curious about the traits of the men she'd slept with and was not particularly interested in the fact that twenty years before, she'd had a Caesar salad, followed by scallops wrapped in bacon. I kind of stopped listening, but luckily the recorder was still paying attention. It wasn't until I was in the process of conducting my next-to-last interview that the existence of courtship feeding in humans suddenly presented itself in a way that was impossible to ignore. I'd been taking the sexual history of a woman who professed to be madly and forever in love with her husband, a man she insisted was the sexiest man she'd ever known. Naturally, I wanted to know what traits had earned him that title. To her great embarrassment, she was unable to list a single one. It was at the end of a long day. I was about ready to wrap up the interviewing phase of my project. Perhaps feeling a bit too empathetic, I tried to coach her out of her dilemma. So I asked her to think about something he'd recently said or done that had triggered the rush of lust she credited him with being able to inspire. She held her head in her hands, scrunched her eyes tightly shut, took a deep breath, sighed, then suddenly sat bolt upright and opened her eyes in astonishment. "I know," she said. "I remember. He brought me two pineapples. He brought me two fresh pineapples!" And at that moment, a mental image of a male lizard scampering up to his intended with a bug in his mouth made *me* sit up and take notice. Courtship feeding. Of course. I was glad I had word-for-word typed transcripts. I went home that night and dragged out those stacks and piles of dog-eared, color-coded papers. Neglected descriptions of food were everywhere. It wasn't the fact of a meal shared. The important thing was the *timing* of those meals. In almost every case, the remembered gift of food had happened just before the couple had sex for the first time.

We see courtship evidence of courtship feeding in a number of species. Now I've documented its existence in humans, I can't help but consider the evolutionary role it might have played in driving recent changes in *Homo sapiens*. The Bajau people of Southeast Asia are such exceptionally skilled divers that in 2015 they attracted the attention of Melissa Ilardo, at that time a graduate student in genetics at the University of Copenhagen. She'd heard stories about how long the men in this community of spearfishers and oyster divers were able to stay underwater—plunging more than two hundred feet down using no specialized equipment other than wooden goggles—and wondered if their skill might have a biological correlate. Could centuries of a lifestyle dependent on diving have driven the evolution of changes in the spleen, an organ known to play a part in the diving reflex of marine animals such as seals, but one whose relation to that reflex in humans had not been studied? Using ultrasound scans, Dr. Ilardo compared the spleen size of the Bajau to that of the Saluan, a closely related community of farmers living only fifteen miles inland. Bajau spleens were fifty percent bigger than those of the Saluan. So far, so good, but the simple fact of larger spleens was not proof of evolutionary change. The increase in size could simply reflect the lifestyle of these people, in much the same way that identical twins having the same genes regulating muscle mass would differ in appearance if one twin were more diligent than his brother about regularly lifting weights. However, even before Dr. Ilardo initiated her genetic studies, there were hints that the differences might reflect more than simple patterns of use. Not all Bajau people are divers, but even those who earned their livings in other ways exhibited unusually large spleens.

The search was on for the genes that made it possible for this boat-dwelling community of divers to stay underwater in ways so far beyond what we think of as humanly possible that they had impressed even scientists living a great distance from their native waters. After all, Dr. Ilardo learned of them when she was still working in Denmark. Given that, it's no surprise that their feats were legendary among the other islanders in the region. Rodney C. Jubilado, a University of Hawaii anthropologist, first became aware of the Bajau while growing up on a nearby island in the Philippines. The article in the *New York Times* quoted him. "We were so fascinated that they could stay underwater much longer than us local islanders," Dr. Jubilado said. "I could see them literally walking under the sea."

What enables humans to walk under the sea? After analyzing cheek swabs taken from the Bajau, the Saluan, and some other populations in New Guinea and China, Dr. Ilardo and Rasmus Nielsen, a geneticist from the University of California, Berkeley, and joint author of a paper on the subject they published in *Cell*, concluded that the evidence points, at least in part, to the frequency of an already existing gene called PDE10A. There are more alleles carrying the variant supporting diving abilities among the Bajau than are to be found in other groups of people. How long ago did the differences that separate these populations occur? Some say that the selection pressure could have happened over eleven thousand years ago at the end of the Ice Age when rising sea levels turned that region of Indonesia into a series of islands. This hypothesis appeals to those who subscribe to the conventional Darwinian wisdom that evolution requires an expansive period of time in which to enact these kinds of changes, and while it might explain why the Bajau evolved large spleens, it doesn't offer much insight into why those living on surrounding islands did not. Given recent research on snail kites, *Anolis* lizards, etc., and my own bias in favor of the ability of a combination of natural and sexual selection to execute fairly rapid adaptations, I guess it's no surprise that I am more drawn to an explanation suggesting that the opening of trade with the Chinese in the sixteen hundreds drove the market value of sea cucumbers high enough that the risk involved in harvesting them from the deep ocean floor was offset by the price that could be commanded. At that point, any slight difference in spleen size that enabled one diver to out-compete his fellows would have served as a pretty strong selection pressure for the unusual spleens that now set the Bajau apart. Geographic isolation in this relatively small population would have favored adaptation of any variant conferring an advantage in diving ability. After all, the Hardy-Weinberg equation, a mathematical calculation of changes in gene frequency from one generation to the next, proves how quickly evolution can happen in small isolated populations with nonrandom mating. It's easy to grasp the concept of small, isolated populations, but just what drives assortative or nonrandom mating? Sexual selection is the most likely and obvious answer. The Hardy-Weinberg definition of evolution comes to us in the form of an algebraic formula developed independently in 1908 by the two men whose name it bears. The actual equation knocks many students of population genetics off balance by demonstrating just how very *difficult* it is for evolutionary change to happen in large populations with random mating. Death rate control of phenotypic variations in things such as coat color doesn't result in the sorts of evolutionary change that conventional wisdom had led most of us to believe would happen. It's only when the data are tweaked in such a way that the populations are neither large nor characterized by random mating that evolution occurs. In other words, it's in small archipelagic populations, such as one finds in the Galápagos Islands, where researchers might want to start looking for sexual selection as the missing third ingredient responsible for any rapid, measurable, and relatively stable change they document. Hardy-Weinberg makes a point that we'd do well to remember: evolution doesn't require dramatic mutations leading to speciation events. After all, the diving abilities of the Bajau don't mark them as a separate species but simply as family members with some extraordinary talents. Neither does evolution require many thousands of years. All it takes is simple changes in gene frequency from one generation to the next within a breeding population. The Bajau live in small, isolated island communities. Throw in a bit of sexual selection, and the stage is set for a demonstration of just how quickly evolution can accomplish its task.

When I first read accounts in the *New York Times* describing the astonishing diving ability of these people, I was struck by the way in which the scientists involved tried to explain how PDE10A became so common in the Bajau. Rasmus Nielsen summed it up in a single sentence, "That's the question that's harder than others." Melissa Ilardo, the original researcher on this project and one of the geneticists working on the analysis of the cheek swab data, was more willing to speculate that natural selection would have favored the variant in question because deep sea diving is so dangerous, "I would think, as morbid as it is, that if they didn't have this, it would kill them."

But what if nobody had to die before his time? What if all it took to unlock the riddle of a people who can walk under the sea was to stir a bit of sexual selection into the mix? I can't quite shake the image of a snail kite offering the object of his desire, a fat mollusk. However, photographs I've seen of sea cucumbers—soft, slug-like things of warty hide—make it difficult to imagine that they could embody the erotic enticement of a gift of chocolates or two fresh pineapples, but what if, *what if*, the ability of a diver to harvest a disproportionately large quantity of this valuable commodity translated into an increase in his material wealth of another sort? I've asked many women what made a man sexy, only to have them respond by describing the high-end automobile the gentleman in question drove. Is it too much of a stretch to speculate that a rise in the price commanded by sea cucumbers in the Chinese marketplace could have been accompanied by a rise in the price commanded in the reproductive marketplace by the diver most skilled at procuring them?

Even in cultures in which marriages are arranged, young people find ways to communicate their desires. In humans, it isn't only the *reproductive* generation involved in deciding who marries whom. Bride price demands are often made by the parental generation, but it is still a reproductive marketplace in which the resource accruing ability of the young male and/or his family is assessed. Sexual selection driven by female choice is in operation whenever closely related female relatives—usually aunts and mothers—influence the decision-making process announced and executed by males with formal control. Actual control can be quite a different matter. Women watching the popular movie, *My Big Fat Greek Wedding,* laugh knowingly at the lines with which a mother reassures her daughter, who has just protested the futility of going against the dictates of her father, the head of the household. "Let me tell you something, Toula. The man is the head, but the woman is the neck. And she can turn the head any way she wants."

It's not always easy to figure out who is really in charge. This is true even in cultures where courtship practices are largely enacted by the young men and women of the reproductive generation. The amount of influence women have in orchestrating those choices is rarely fully credited by anyone. It's not just among humans that this is the case. Researchers who study sexual selection in other species and who readily admit that this evolutionary mechanism is driven by female choice, have a great deal of difficulty reporting just what it is that females are doing. In the opening pages of Michael Ryan's book, *A Taste for the Beautiful,* he describes observing the Resplendent Quetzal and informs us that birders travel from the far corners of the earth to Central America to observe what has been called the world's most beautiful bird. Ryan admits that his descriptions apply only to the male, which he compares to a brightly decorated Mexican piñata. He tells us, "I also saw a female quetzal, but no matter. She lacked all the male's fancy embellishments, and I hardly gave her a second look."

I don't blame him. Anyone who has heard lectures by Richard Prum, author of *The Evolution of Beauty: How Darwin's Forgotten Theory of Mate Choice Shapes the Animal World–and Us,* and seen his videos of the courtship displays of birds deep in the tropical jungles would be equally hard-pressed to describe what the females were doing. I consider myself a student of courtship behavior. I'm really curious about the ways in which females go about selecting their mates. Knowing that female behavior flies beneath the radar almost by design,

I like to think that I've trained myself to pick up on things that others just might not notice. Matters of appearance are the easiest data to record. More difficult are subtle behavioral cues and strategies. I've seen the video of five adult male Blue Manakins out on the branch that serves as their stage performing synchronized, cooperative displays for the benefit of a visiting female. Ask me about her appearance. My response would echo Michael Ryan's. "I hardly gave her a second look." It's almost impossible to pay even a token amount of attention to the audience when all that leapfrogging and jumping is being executed in rhythm by flamboyantly colored males decked out in electric blue feathers, with accenting collars of ebony and caps of brilliant red. What did the female look like? I have no idea. What was she doing? I can only assume that she was as deeply enthralled with the exuberant demonstration of excess and artistry as I was. But maybe not. Prum reports that in many manakin species, females don't even witness most of the displays, but they nonetheless choose their reproductive partners from the alpha males in charge of the show. We've all heard the expression, "Beauty is as beauty does." However, these guys don't appear to be doing much of anything other than being almost inconceivably beautiful. That's not to say they don't put in long hours practicing their art and waiting even longer years before it's their turn to take center stage. After all, they don't even get to rank as understudies until their plumage has assumed full adult magnificence at about four years. Ornithologists willing to trek into the Brazilian jungle can *see* what the males are doing. The question for me is *why* female Blue Manakins are willing to ask for so little other than an occasional front row seat, even though any aspiring P. T. Barnum might justifiably describe the display as the greatest show on earth. It's not as if the females have nothing to offer in exchange. Their eggs are a very valuable scarce resource. Not only their eggs but also the labor that goes into nest-building, laying, incubating, and feeding the young. Male manakins help with none of this. Once fertilization has occurred, the females are on their own.

In manakin species, females exert a strong selection pressure not just for beauty but also for an obligatory male-male cooperation that Prum tells us is a total loss for most males. In one species in which the DNA of the chicks has been compared to that of the displaying males, 90 percent of the males failed to father any offspring. Why is a female willing to work so hard all by herself to raise male offspring when so few of them will get a chance to carry her 50 percent share of his genetic material into the reproductive generations of the future? A manakin's daughters may not grow up to be as pretty as her sons, but female chicks are a much better bet for the long-term survival of her genes. It's a complete puzzle to me. I can't help but think back to widowbirds. How did they earn that name? Was the moniker nothing more than speculation by early naturalists, or did someone actually witness these males being cut down in their prime? I can find descriptions of predations of widowbird *nests*, but not any research documenting the disproportionate death rate of adult males. And how about the manakins? It seems logical that predators in these jungle habitats would have figured out that wildly colored, displaying males make much easier targets than the less conspicuous females. But, once again, the only data I can find describe nest predations, not the disappearance of live adult birds into the jaws or talons of another creature. Is that because stationary nests are easier to film with hidden cameras? That makes no sense. After all, the courting sites, known as leks, are well-established, and there is lots of footage of the males' attention-getting displays.

There are two things that impress me about manakins. The first is the sheer beauty of the males. The second is the total lack of parental investment by these same males. In a species in which that initial donation of sperm is the *only* thing the fathers give the chicks or their mothers, why would a female manakin *care* if she became a widow? Since her reproductive partner already disappears immediately after copulation, she'd have no way of even knowing if he was dead or alive. Gorgeous as they are, perhaps most male manakins are a disposable resource. When a female invests energy in raising a son, she is playing a genetic lottery with the odds stacked against her. Up to 90 percent of male Blue Manakins never even have sex. From an evolutionary viewpoint, they might as well be dead. However, if one of a female's sons does beat the odds and ends up becoming a breeding alpha, the payoff would be enormous, great enough I guess to keep her dropping quarters into the slot machine, eggs into the nest. And should none of her sons grow up to become reproductively active adults, no matter, her daughters will pick up the slack.

I started this whole discussion of manakins not because I wanted to understand why the males were so beautiful but because I wanted to know what the nondescript females might be doing. My first thought—one for which I can find only hints of speculative supporting evidence in a few other species of birds characterized by lek behavior—is that perhaps the very drabness of the females, perhaps their very *invisibility*, protects them from predation and enables them to successfully raise both their male and female offspring. If that were the case, by selecting for males that would conspicuously draw attention to themselves, female manakins would be engaged in *designing* mates that—willingly or otherwise—would sacrifice themselves so that predators would ignore the drably colored female and her equally nondescript offspring. If true, such sacrifice would demonstrate an expression of cooperative behavior, freely given or not, far more impressive than any fancy dance move. Should a permavirginal male be the one picked off while his relatives on the maternal side were spared, he would enhance the inclusive fitness of his family lineage and finally be granted a true shot at genetic glory by contributing to the survival of the genes he shares with his mother, siblings, and other relatives closely linked by DNA. The genetic payoff would be worth all the years he spent waiting his turn to join the chorus with no hope of ever becoming someone's sweetheart and would support the idea that those coordinated routines have to be about more than just the beauty of the dance.

Strategies such as my speculative account picturing female manakins and widowbirds as if they were little more than fly-fishermen choosing the arrangement of feathers most likely to lure predators to their reproductive partners would, if true, fall into the category of evolutionary logic known as inclusive fitness, which turns to degrees of relatedness, or kin selection, as a way of explaining behavior which at first glance appears to be at odds with the self-interest of the individual performing it. Richard Prum considers—and appears to reject—kin selection as the force shaping the cooperative displays of male manakins. He cites a study that tested the theory of inclusive fitness by measuring how closely related to each other the males within displaying partnerships of Long-tailed Manakins were. Genetic tests demonstrated that the males in these groups were not, in fact, any more closely related to one another than would have been predicted by chance alone—all well and good. Prum convinces me that the degree of relatedness within a performing team of males in this population cannot explain the evolution of their rather bizarre cooperative behavior. However, I can't rule out inclusive fitness as an explanation for the displaying behavior of manakins until someone also calculates how closely related *all* the performing artists, not just the breeding alphas, are to the females and the chicks in various nesting sites. Initial studies documenting a correlation between the location of the leks where the reproductively active alpha males perform and the position of the nests holding the chicks they fathered intrigue me. Why would a female Blue-crowned Manakin not build her nest close to the lek from which she has chosen her reproductive partner? Because as it turns out, the females of this species *don't* go for proximity. They select their mates from the *third*-nearest lek. I want someone to figure out how that behavior might benefit a female's chicks. I'd also like to see more data on male to female sex ratios in manakins.

I just can't bring myself to let go of the widowbird aspect of it all. Because the thing is this: much as I believe that female birds select for beauty, I have a really hard time believing that females can afford to choose for beauty *alone*. The potential costs are far too high. Females of all species have to make selections with an eye to ensuring that whatever it is those males are doing, it's going to in some way contribute to her ability to get her own genes and those of her offspring into the gene pool of the next reproductive generation. It's time to force ourselves to carefully observe female behavior in all species. Prum mentions research by Gail Patricelli, who began studying Satin Bowerbirds when she was a graduate student working with Gerry Borgia. Rather like my going back through the typed transcripts of conversations with women to discover things I'd ignored in the course of conducting the interviews, when Patricelli studied the films they'd made of the courtship encounters between a male bowerbird and the female he was inviting in to see his etchings, she documented subtle changes in the postures assumed by female bowerbirds. A female's signals of wariness in the face of overly aggressive male displays caused him to modulate his own behavior. Subsequent experiments confirmed her hypothesis that male bowerbirds who were able to regulate the intensity of their display in response to cues from the female were the most successful at attracting mates. If we really want to understand how evolution works, we've got to start paying closer attention to what females are doing. Males of all species with their displays and blusters, big antlers, loud voices, and confident struts and gestures make it very difficult to pay much attention to females at all, whether in a rain forest or on the Senate floor. This is especially true when one is attempting to riddle out the comparatively subtle glances and movements that signal interest, encouragement, and intent between those involved in the courtship dances of sexually reproducing species.

Humans are the only species in which I've done any empirical research. As a member of the species I'm studying, I have the advantage of understanding the meaning of the signals we share. However, as I'll note in my subsequent chapter on the evolution of language, words can be used to reveal and conceal in equal measure. When investigating reproductive behavior, recording actual outcomes—such as whether the couple had sex, got married, used contraceptives, had children—is important. But of course, even then, I am relying on what someone *says* she has done. Memory and language can both enhance and complicate data collection. When a woman remarked that a man was handsome—in other words, when she claimed to be selecting for beauty just as male researchers assume females do—did the physical description of the man match up with what she may have volunteered about her ideal type? Not usually. What appears to be going on with human females is that they find beauty where they have first seen other attributes such as intelligence, control of social and material resources, generosity, confidence, courtship feeding, skill with children, the affirmation of her choice by parents and friends, etc. If I could sum up the differences between the sixty-six women I interviewed and the ways in which men respond when I ask the same question, it would be that women come to see the men they've *come to love* as handsome, whereas men come to love the women they have first perceived as beautiful.

Understanding the relatively rapid evolution of humans is one of the puzzles I've set before myself. We've been around for only a little over 300,000 years and are a young species. Compared to much of the rest of the teeming life on this planet, we haven't stood the test of time, and unless we alter some of the behavior that is damaging the ecological systems on which our own and all life depends, we aren't likely to. The event or events that led to changes in the gene frequencies of whatever genes set us apart from our nearest primate relatives occurred early on in our evolutionary history. Although we continue to evolve in small ways, as evidenced by the diving abilities of the Bajau, it didn't require millions of years to get us *started* down the path that led to where we are today. The crucial moment could have been a rather abrupt one. Some evolutionary pressures simply result in differences in gene frequencies between generations and can produce a population characterized by traits such as those relating to height, hair color, or skin pigmentation. Evolution in snail kites has not, or at least not yet, resulted in a population of birds in which the offspring are dissimilar enough from their parents or grandparents to be declared a separate species. Other evolutionary incidents *are* more obviously dramatic and result in the kinds of speciation events that produce branching in the family tree. While small changes such as those contributing to diving ability in the Bajau are evolutionary changes that can serve as markers setting populations apart in ways that companies such as Ancestry.com or 23andMe can unravel, not every evolutionary change results in a speciation event. This is all complicated by the fact that definitions of what constitutes a species vary. An easy definition refers to a group of organisms having common characteristics and capable of mating with one another to produce fertile offspring. The difficulty here is that this definition is a human invention and the organisms we study don't always behave in ways that make classification easy. Sometimes the reproductive barriers are physical ones that never result in viable offspring or else in hybrids that are themselves incapable of reproducing, but sometimes the barriers are behavioral ones, such as when a male bird sings the wrong courtship song. Humans are both physically and behaviorally a distinct species from non-human primates. Did this happen gradually over time, or was there a sudden mutation that established a clear line of demarcation between us and our nearest primate relatives? We now know that our ancient ancestors mated with both Denisovans and Neanderthals. However, the genetic signature left by those encounters constitutes but a small part of the DNA of modern humans. At some point, the other two populations of early humans died out and we lived on. What was going on?

We share over 97 percent of our genetic material with chimpanzees and bonobos, making them our closest primate relatives. Sometimes an apparently small change in chromosomal structure can result in widespread differences in terms of how existing genes are expressed. Not every single feature that sets species apart needs to have been subjected to individual selection pressures. Roughly seven million years ago, our ancestors split off from the primate line that led to chimps and bonobos, Great Apes that are characterized by 24 pairs of chromosomes. Modern humans, Neanderthals, and Denisovans all have 23 pairs. Our path diverged from the Great Apes when two of their chromosomes fused into the one that is now ours. This seems a small enough difference—23 pairs of chromosomes for us, 24 pairs for the Great Apes—but it is exactly such seemingly minor shifts in genetic architecture that can result in fairly major changes in the ways in which genes are expressed across the board and throughout the organism.

When trying to understand exactly how evolution works, I tend to focus on two important areas that Darwin either neglected or misunderstood. The first is the role of sexual selection in shaping all kinds of traits, not just those relating to beauty or secondary sex characteristics. The second is the amount of time required. One of the most fascinating examples of evolutionary speed is documented in research conducted by Jonathan Losos. In his recent book, *Improbable Destinies, Fate, Chance, and the Future of Evolution*, Losos mentions sexual selection almost in passing but makes a compelling case for the rate at which evolution can happen. The introduction immediately captured and focused my attention with these sentences: "It turns out that Darwin and a century of biologists following him were wrong in one key aspect: evolution does not always plod along at a snail's pace. When natural selection is strong — as occurs when conditions change — evolution can rip along at light speed." I'd been waiting years to hear a well-respected biologist make such a bold statement. Losos gives examples of rapid change in a number of species. Of particular interest to me was a study that Losos himself conducted when he released anole lizards on Caribbean islands that lacked existing representatives of this genus. What happened next surprised everyone with its demonstration of just how quickly changes in the appearance of these lizards happened. It was easy to tell one generation from the next just by looking. Losos turns to adaptive radiation and convergent evolution in order to explain the mechanism behind the drama. Whatever the driving force responsible for the phenomenon, Losos and his lizards convincingly challenge a rather basic Darwinian assumption regarding the slow, steady nature of change driven by natural selection. Why do we assume that only *natural* selection is at work here? There might be more to this story than initially meets the eye. Early research on sexual selection in anole lizards postulated that females of the species were skilled at appraising just which males were best at exploiting the resources necessary for survival. Could something like that have played a part in sculpting the adaptations that happened at light speed in the lizards released on those Caribbean Islands? Let's take a look at it all from a female point of view.

Chapter 3: Moving Right Along

I can understand why Jonathan Losos was surprised by the rate of speed with which his lizards evolved. After all, most of us have been thoroughly indoctrinated by the Darwinian dogma of an evolutionary process that takes its sweet time. I readily admit that I found my first exposure to the Hardy-Weinberg equation deeply unsettling. After attempting to solve the problems presented at the end of each of the first few chapters in my population genetics textbook, I threw crumpled-up page after crumpled-up page of incomprehensible solutions across the room. Expecting to see predation-driven changes in the equilibrium of various genotypes in large random mating populations and not finding them, I was certain I had forgotten whatever knowledge I'd once had of algebra. My professor came to my rescue by demonstrating the power of sexual selection in small isolated populations to produce the kinds of changes I'd been certain could be achieved in large panmictic populations by nothing more than hawks preying on white rabbits cursed with pairs of recessive alleles for coat color. I learned the hard way about the power of sexual selection to effect *rapid* change, but I pushed the insight aside and didn't really begin to understand just how much of a force to be reckoned with it was until I sought to understand the specifics of human evolution.

Since Darwin drew so heavily on examples from animal husbandry when making the case for evolution in *The Origin of Species by Means of Natural Selection,* I remain puzzled that he failed to internalize and integrate his firsthand knowledge of how quickly changes *within* a species could happen when he later wrote about human evolution in *The Descent of Man and Selection in Relation to Sex.* In his defense, he was up against some pretty deeply ingrained religious beliefs that stated categorically that all species were created in their present form by God. Before tackling the difficult task of convincing people that one species could evolve into another, Darwin first hooked his audience by turning to the British affinity for animal husbandry and the skill with which backyard hobbyists had created exotic breeds of everything from pigeons to dogs, cats, sheep, cattle, horses, and so many other fanciful varieties of animals as to require an entire fleet of Arks should another flood threaten to cover the earth. Darwin never claimed that these various breeds of pets and farm animals were distinct species. After all, the only reproductive barriers between a Collie mating with a Corgi were those imposed, often with great difficulty, by their human owners! However, by using examples near and dear to the hearts of his fellow countrymen and women, he effectively demonstrated that the specialized biological changes that adapted various creatures to specific niches, both in terms of appearance and behavior, were not only possible but could also happen fairly quickly. However, as far as Darwin was concerned, this nod to the power of selective breeding was not intended to prove anything other than the possibility of morphological change. He was not interested in pointing out the ways in which mating decisions impacted the *rate* of change. Perhaps he didn't even notice.

When Jonathan Losos first began investigating the adaptive properties that set different species of anole lizards apart, his goal was not to upend conventional wisdom regarding evolutionary speed. However, when these adaptations happened much more quickly than he had anticipated, he, unlike Darwin, did take notice. Losos's adventures in evolutionary biology got a jump start when he was still a sophomore in college. One of the graduate students asked Losos to go with him to Jamaica to assist in a field study of *Anolis* lizards. They looked primarily at four species: the Jamaican bush anole, Graham's anole, Garman's anole (known to Jamaicans as the green guana), and the Jamaican twig anole. The lizards were immediately distinctive from each other in a number of regards—size, shape, coloration, leg length, toe pad adhesion, and dewlap display—that adapted them to their particular niche in the habitat. Dewlaps are employed in communicating territorial and reproductive intent. Leg length is niche dependent and enhances both hunting prowess and predator avoidance. The students took some of these lizards back to the lab, measured leg length, and correlated it to the skills needed in the neighborhood where that particular lizard lived and hunted. There were distinct differences in leg length between the bush anole and the twig anole. I see this early research as being a fortuitous start to the interest in leg length that later provided Losos with the aha! moments at the heart of his discovery that "evolution can rip along at light speed."

Lizards that live on small landmasses surrounded by water look so much like their cousins occupying similar niches in the ecosystems of geographically equivalent islands that it's easy to assume that the doppelgangers are closely related. Easy, but wrong. Despite being nearly identical in their adaptive colorations, size, leg length, hunting behaviors, toe pads, etc., the degree of relatedness between lizards who occupy similar vertical real estate on analogous trees on nearby islands is more apparent than real. Genetic analysis fails to support a case for common ancestry between these look-alike-lizards and instead makes a pretty convincing argument for adaptive radiation. The lizards on any one of the islands studied are more closely related to each other than they are to their body doubles on the neighboring islands. But while convergent evolution and adaptive radiation are pieces of the puzzle that one must consider when attempting to unravel the mystery of evolutionary modification, and while they are no doubt factors in explaining the speed that Losos eventually documents, later in this chapter, I'll argue that the role of sexual selection should also be part any investigation evaluating how organisms survive and change in new environments or in old ones that have suddenly altered. However, before doing so, let's spend a bit of time with Losos and the biologists with whom he worked and jointly produced a paper on evolutionary speed in *Nature*. The journal described their article in a press release, "This may be among the most important work in evolutionary studies since Darwin studied the finches on the Galápagos Islands during the voyage of the Beagle." Needless to say, the *New York Times*, the *Boston Globe*, *USA Today*, and *ABC News* were all over the story, briefly turning a remote island in the Bahamas into a media circus. Quite a bit of research with anole lizards, beyond what I've described so far, had taken place in the intervening years between Losos's undergraduate measures of their leg-length and the commotion that he modestly describes as his fifteen minutes of fame.

Losos became fascinated by the idea of adaptive radiation and was particularly taken with the research conducted by a pair of husband and wife biologists, Tom and Amy Schoener, who had spent years studying Caribbean anoles. While finishing up his dissertation, Losos read a paper they had written on the replicated adaptive radiation of *Anolis*. He contacted them and was invited to join their lab at the University of California, Davis. Losos eventually ended up on Staniel Cay, a small island in the Bahamas, where he continued the work the husband and wife team had begun. His job was to capture as many lizards as he could from the fourteen island populations established by the Schoeners at least ten years previously. Once again, Losos found himself measuring legs. After a series of frustrations and adventures of the kind so often associated with fieldwork of this type, Losos summed up the results in the paper that led to all the aforementioned media excitement in a single sentence: "The population had not only evolved but had done so exactly as we'd predicted: on islands where lizards used thin branches, they tended to have very short legs, whereas, on islands where they used broader perches, their legs generally were longer." Since these changes had occurred in no more than a decade, a case for rapid evolution had been effectively made.

That stunning conclusion led to a subsequent field experiment in which evolutionary speed and adaption were tested in response to predation. This time, control islands onto which humans had released no new species were compared to experimental islands where the scientists introduced large, predatory curly-tailed lizards twice the length and ten times the weight of the smaller brown anoles already living there. After only six months, anole populations on the experimental islands had half as many lizards as could be found on the control islands. After only two years, brown anoles on curly-tail islands had not only developed shorter legs but had also used those legs to scramble out of the way of their predators and move up into the narrower twigs found near the top of the shrub-like vegetation found in the geographies of both the control and experimental groups. Hurricanes ended that experiment after only two years, but the results were tantalizing enough that Losos took advantage of the blank-slate opportunities provided by hurricanes to design an experiment that would further challenge the conventional wisdom found in Darwin's understanding of evolutionary speed.

Losos experimentally introduced a species of brown anole lizard onto tiny Caribbean islands that had been washed clean of any sort of lizard population by hurricanes. He was gathering more evidence in support of rapid evolution and was not at all interested in imposing his will on that of the lizards by dictating mate choice. Well, that is not completely accurate. He did, after all, populate each island with but a single mating pair. Not until those two got around to doing some egg laying of their own was there much of a choice at all. Once Losos released his foundation stock, he was quite willing to let nature take its course and give natural selection a chance to come to the rescue of lizards that had lived in the upper reaches of tall trees with thicker branches than those found on the narrow twigs typical of the bush-like growth in their suddenly unfamiliar surroundings. The leg lengths with which they were equipped when they arrived were not nearly as functional in this new habitat as they'd been in the one in which they were captured. Losos repeated the procedure of introducing a single male-female pair on each of seven neighboring islands that, thanks to hurricanes, lacked lizard populations of any sort. At the end of a year, the mating pairs had survived. When Losos and company returned after two years, things were moving in the expected direction. Leg lengths were shortening, and they were doing so even faster than had been anticipated. *In only two generations*, the legs were measurably shorter, a trend that continued over the next four years. Six years is a long time for these tiny islands to maintain themselves as living laboratories. Hurricanes have an inconvenient habit of doing periodic house-cleaning and data collection did not continue beyond that point.

Given a foundation stock consisting of only a single mating pair, it goes without saying that all subsequent generations of these lizards are going to be closely related. Here's a thought experiment. Instead of thinking of this exercise as a demonstration of the power of *natural* selection to quickly adapt a species to a changed environment, try thinking of it as an example akin to *artificial* selection. Imagine a sportsman whose favorite gun dog suddenly displays an unexpected talent when retrieving a pheasant that his owner has just brought down. Because the hunt has taken place in a field that contains a small pond and the bird happened to be flying over this pond when it was shot, instead of hitting the ground, the body of the pheasant plops into the water. This breed of imaginary dog refuses to make water retrievals. However, this *particular* dog has never been informed that his kind won't go near the water and, after just a few moments of anxious pacing on the shore, he plunges in and comes back with the bird in his mouth. His owner could not be more delighted with his favorite companion's newly demonstrated talent. Because his property is spotted with a series of ponds, he wishes all his dogs were as adept. When puzzling over what might have made this dog so uncharacteristically willing to splash in, the hunter recalls that the dog's mother has never minded stepping out into the rain and getting her feet wet. The next time she comes into heat, the only stud that he allows near her is her water-retrieving son. From the resulting litter of pups, he notes which ones have a habit of smacking their tiny paws into the water dish and will keep those youngsters and make them part of his foundation stock. People who practice animal husbandry refer to this not as incest but as linebreeding. In only a couple of generations, linebreeding for a selected trait can result in a family of retrievers that are as comfortable jumping into the water after a fallen duck as they are racing through the brush to bring back a pheasant whose life has been cut short by a hunter with deadly aim. In the blink of an evolutionary eye, our gamekeeper has designed a duel-purpose breed that, as long as he makes sure to equip his kennel with fences strong enough to prevent accidental matings, will eventually breed true.

Now let's think of ways to compare the hypothetical situation above to the reproductive behavior of the anoles on a Losos island. There won't be any linebreeding possible in the first lizard generation. Things won't really get started until the first eggs have been laid and have hatched and the tiny lizards have grown to reproductive maturity. At that point, there is going to have to be, if not a little Oedipal indulgence, then perhaps some Electra-like enchantment, or at the very least a few sibling relationships characterized by something other than rivalry. Because the gene pool is so small, true random mating is not possible. Stay with me. I know this does not yet conform in its entirety to my hypothetical example of linebreeding. After all, we're not selecting for a specific trait such as water retrieving in a breed of dogs that had not previously shown any inclination in this direction. There is no all-powerful "other" imposing its will on these lonely lizards, but I think we need to at least consider the possibility that the lizards may have a few opinions of their own about who should be mating with whom. Sexual selection was not the focus of this study. Natural selection was, and it is easy to provide an explanation for the observed changes that rely on natural selection alone. Suppose that lizards with slightly shorter legs were better predators than their brothers and sisters with the old-fashioned length. It's possible that their superior skill at feeding themselves might enable them to survive while last year's leggy models died out—natural selection at work. But be willing to consider that the lizards themselves might also be doing an additional something not too far removed from the hypothetical example from linebreeding that I've sketched out above.

Although Losos and I may differ in our weighting of the factors that explain the relative speed of evolution in some populations, we agree that evolution can, at least in certain circumstances, happen pretty quickly. What about Darwin? Why didn't he see that what he called selective breeding is actually a form of sexual selection, albeit one in which the actor making the choices is from outside the species being modified? Think of it as sexual selection by proxy. Viewed objectively, selective breeding is really nothing more than a rather interesting form of what Hardy-Weinberg called assortative mating and what I call sexual selection. It might help to recall the distinction I made in Chapter One between the two evolutionary forces: Sexual selection determines *who* gets born; natural selection determines *what* happens to them once they embark on the struggle to survive. Modern biologists refer to the pressures exerted by humans in shaping the various creatures of the earth to our purposes as artificial selection. Human interference in the form of gene splicing, etc., aside, the main thing *artificial* about it is *who* gets to make the reproductive choices determining which individuals are born. In the case of artificial selection, we've outsourced the decision-making process to a member of the species other than the one being refigured. It may work for us, but the changes thus engineered are not necessarily what the species in question would have found to be adaptive. This is especially true when humans select for something such as weight gain in industrial farm animals so rapid and extreme that chickens expressing the trait are unable to stand, could never survive in anything resembling the wild, and have been designed for nothing so much as the kind of fitness best suited to a roasting pan in a 350 degree oven. Just imagine how different those same chickens might look if the power of sexual selection could somehow be restored to its rightful owner, the hen in some ancestral flock. Yes, the *hen*.

When observing humans in all kinds of social settings and listening to them talk about the almost-immediate appeal of one individual over another in the sort of instant connection we refer to as love-at-first-sight, I long ago coined a phrase to explain it to myself. What some might simply call chemistry, I dubbed "the gene recognition theory of attraction" and decided that at least in some instances, the man whose face held the most appeal for a woman bore a passing resemblance to the one found in her own mirror. People practice assortative mating (read sexual selection) in a number of categories, not just appearance, but for the moment, let's ignore cultural measures and stick with nothing other than how we look and move. Humans engage in phenotypically expressed assortative mating. Probably because of that, married couples are more genetically similar to each other than are randomly selected pairs of humans. Anecdotal tales of brothers and sisters who have been separated at birth and who later—unaware of the biological link between them—meet as adults often fall in love. Those stories are the stuff of headlines and whenever I come across one, I can't help but compare them to other news reports documenting the outsized popularity of some sperm donors. I worry enough about these sorts of things that I once suggested—only partly in jest—that babies who are conceived via choices made in the descriptive profiles provided by sperm banks ought to get the number of their particular donor tattooed on their heels. Since many of these kids will grow up in the same geographic area, I predict that they will have a better than average chance of having a love-at-first-sight moment should they, as adults, accidentally lock eyes across a crowded room with an unidentified half-sibling.

A recent essay in *The New York Times* caught my attention with the headline, "First I Met My Children, Then My Girlfriend. They're Related." Written by a sperm donor describing his 23andMe-assisted connection with the first of six children he had fathered in the 1990s while supplementing his cab driver's income with twice-weekly visits to a sperm bank, the author calculates the number of samples he contributed against the odds of conception and speculates that he may have fathered as many as sixty-seven children. What was his reaction to first meeting a few of these kids? "I may be biased, but I found my children to be ridiculously attractive. I felt a sudden need to share their photos with all the ex-girlfriends who chose not to marry and procreate with me." Since I argue for a system of sexual selection powered by female choice, I'm glad he included a description of the way the mother of two of these youngsters responded to him when they first met. The couple, who are now romantically paired, obviously have no genetic link between them. However, he has a pretty powerful resemblance—and genetic connection—to the people she loves most in the world, her daughters. She explained the immediate attraction she felt towards him by speculating that it was due to the mannerisms he shares with her—and his—children. We tend to have an affinity for traits that mark us as members of the same tribe. There is an exception to this general rule. Much as women are attracted to those who bear some resemblance to themselves or to their close relatives, they also can be snapped to attention by the obvious outsider, the man who is immediately set apart by traits that mark him as novel, interesting, and completely worthy of their curiosity and focus.

Why don't children who have been reared together fall prey to the gene-recognition-pull of erotic desire that gives me pause whenever I speculate about the sexual sparks that might fly if two of the potentially 67 half-siblings fathered by that income-supplementing cab driver were to accidentally meet? What is the origin of incest avoidance in humans? It's not something that needs to be taught but is instead something that happens almost automatically in children who are reared together during a critical period. Evidence gathered in a study of kids socialized in collective peer groups in the kibbutz indicates that this period of negative imprinting occurs between birth and age six. The parents of children living in these communal nurseries would obviously be thrilled if their sons and daughters, upon reaching maturity, fell in love and married. However, it doesn't happen. Although genetically unrelated, these little boys and girls spend their early years engaged in as much touching, hugging, and rough-and-tumble contact as would be typical of biological brothers and sisters living in nuclear families separate from the kibbutz. Siblings play and tussle with each other and compete for their parents' attention, but if they've been reared together, they don't respond to each other with romantic intent. What about the kids who grow up in communal peer groups? Joseph Shepher, author of the study on incest avoidance and negative imprinting, reports that "among 2769 marriages contracted by second generation adults in all kibbutzim, there were no cases of intra-peer group marriage."

Not many biologists are willing to grant non-human species as much agency as they assume humans possess. But are *we humans* intelligent enough to understand just how discerning other species are when it comes to making decisions that impact their own lives? Frans de Waal, primatologist, ethologist, and author of *Are We Smart Enough to Know How Smart Animals Are?* doubts that we are. De Waal is one of the few biologists who appears to have no trouble granting instrumentality to species other than our own. I haven't discussed this with de Waal, but knowing the respect he has for the level of intelligence found in other species, I'm willing to bet that he would grant lizards the same ability to identify those who are closely related to themselves as we humans have. Not only that, but I think he'd also be quick to give female lizards credit for the ability to recognize the ways in which shorter leg length might contribute to the resource-accruing potential of male lizards being evaluated as reproductive partners and to choose accordingly. I'm willing to make the assumption that female lizards are every bit as rational as I am but don't expect most others to go out on that limb with me without further research. Granted, those who find themselves taking exception to the theories

I'm developing here may feel that my own conceptual abilities have not risen above those lodged in the reptilian brain, but no one who has ever met either my first or second husband could possibly challenge my skill at mate selection.

Here's what I'm proposing: I think the rapid evolution that Losos documents in his island lizards is due to a *combination* of natural and sexual selection. Jonathan and I have emailed back and forth about these things. So far, he remains unconvinced that sexual selection has played a part in the evolutionary speed he substantiates. He points to an example in his book that demonstrates adaptive change in an *asexually* reproducing species, the common bacterium, *E. coli*. Microbiologist Rich Lenski is known for understanding the importance of bringing the experimental method that is the foundation of good science to studies of evolution. Lenski set up a laboratory experiment designed to explore the evolutionary process using microscopic organisms that reproduce *asexually* and *fast*. When food is readily available, these tiny single-celled organisms can divide as quickly as *every twenty minutes*. Now that's what I'd call a mind-blowing, flash-in-the-pan generational magic trick. At the risk of glossing over the impressive complexity and precision of Lenski's experiment, I will cut to the chase. After only six years, there were statistically significant differences in the evolutionary adaptations found in his experimental and control groups. Since we're used to thinking of evolutionary change that takes place over many thousands or millions of years, this would seem to reinforce the data generated by those headline-grabbing lizards and make the case for natural selection's ability to produce really rapid evolution. Losos points out that since *E. coli* reproduce asexually, the six-year timespan makes it pretty clear that sexual selection is not a necessary ingredient in any explanation of evolutionary speed. I say, "Not so fast." Those six years reported on changes that occurred in ten thousand generations. As I quipped to Losos, "No self-respecting sexually reproducing species would brag about change that required *ten thousand generations*." The point I want to stress is that when measuring evolutionary speed, the yardstick should not be calibrated by a human understanding of time but needs to take into account the generational time span of the species in question. Time is a relative thing. If an organism lives for only six months, we can't determine evolutionary speed by somehow grounding the measure of its biological experience in a *human* understanding of time. It's the number of generational replications that must be the basis for any true measure of speed. Snail kites: morphological change in two generations. *E. coli*: morphological change in ten thousand generations. And the winner is . . . drum roll, please . . . sexually reproducing species!

My initial focus on the importance of resource-accruing ability as one of the factors human females evaluate when selecting reproductive partners was influenced by a paper on female choice as practiced within, interestingly enough, a population of anole lizards. At the very time I was reviewing the literature on female choice's role in evolution, Robert Trivers, who then was affiliated with Harvard, had written a paper, "Sexual Selection and Resource-Accruing Abilities in *Anolis garmani,*" which was published in the journal, *Evolution*. In the abstract, he wrote, "Meanwhile, several theoretical advances (Maynard Smith, 1971, Williams, 1975, Zahavie, 1975) have convinced me that female choice is a central factor molding male sexual dimorphism and systems of male-male competition. The data presented here for *Anolis garmani* tend to support the view that those males who do most of the mating are those who have tended to demonstrate the superiority of their own non sex-linked genes." Because I was already convinced that Darwin's emphasis on sex-related traits seriously underestimated the power of sexual selection, I knew I had to meet the man who grasped the potential importance of female choice in shaping the expression of non-sex-linked genes. At the time, I was in the early stages of framing the research project that would become the basis for my Ph.D. in anthropology from Rutgers. Ignoring the advice of friends who knew more about automobiles than I did, I climbed into my high-mileage, beat-up old VW bug and drove straight from New Brunswick, New Jersey, to Cambridge, Massachusetts, where I persuaded Dr. Trivers to serve as the outside member on my dissertation committee.

Because only a single egg matures at a time and because female *Anolis garmani* copulate only once in order to fertilize that egg, these lizards are well suited to studies in which a male's reproductive success can be measured simply by counting the number of his copulations. The resource-accruing ability of these same males was calculated by weighing and measuring them. Operating under the logic that males indicate their skill at resource accrual by catching and eating enough insects to enhance their physical size as compared to that of their less-accomplished peers, Trivers demonstrated that female choice for resource-accruing ability was evidenced by their reproductive preference for larger males. The possible relationship between age and size was taken into account and shown not to be a factor. The last few sentences in the article concluded, "What is striking about our data on *A. garmani* is that those males who have maximized growth rates times survival are those who inseminate adult females. Although I have no data on female choice, our other data suggest that females naturally mate with males whose daughters will have the best chance of maximizing resource accrual times survival. That there is a strong system of male-male competition for females does not imply that females are merely mating with males good at besting other males. Instead, systems of male-male competition are expected to evolve under the influence of female choice, so that males compete among themselves in such a way as to reveal to females that their genes are good at maximizing survival times resource accrual."

I'd like to conclude this chapter by heading back to the Galápagos Islands, the geography Darwin made famous with his claim that it was the finches on those very islands that convincingly demonstrated natural selection's ability to drive speciation events. By the time I finally sat down with *The Origin of Species*, I'd already taken the course in population genetics that had had me butting heads with Hardy-Weinberg. Therefore, I was a bit skeptical about Darwin's claim that the finches on the Galápagos Islands offered proof of the power of *natural* selection. I wanted the professor who'd convinced me that it wasn't my math skills that were leading me astray to figure out how to explain a thing or two to Darwin. I wanted Darwin to fully grasp the implications of some of the constraints written into those end-of-chapter problems—population size, degree of geographic isolation, and whether mating behavior was random or selective for certain traits—and consider them in light of his observations regarding the beak size of the finches on the Galápagos Islands. Of course, since Darwin was dead by the time Hardy and Weinberg developed this formula, and since Hardy and Weinberg were themselves both dead at the time I needed them most, and since I'm uncertain of the actuarial status of a professor whose name I can no longer recall, such an encounter would have involved a few things in addition to a time machine. What I really wish had happened was that Darwin and all biologists following him could have understood that the speciation events that were so evident on the Galápagos Islands might be much more efficiently explained by turning to the power of *sexual* selection than they were by relying only on the mutation-dependent canvas of *natural* selection. I wanted Darwin, and all subsequent neo-Darwinists to at least consider that *both* selective pressures might be at work. The fact that the Galápagos example is still considered a gold standard proof of how *natural* selection works continues to frustrate me.

Much as I might want my obituary-status-unknown professor to have a little talk with Mr. Darwin, you and I don't have to do any time-traveling at all in order to witness the birth of a new species of finch on the Galápagos Islands. The November 23, 2017, Science & Environment section of the BBC online news contained an article by science writer Rory Galloway titled: "Galápagos Finches Caught in the Act of Becoming a New Species." In 1981, Professors Rosemary and Peter Grant, who had been tracking population trends on Daphne Major, a small Galápagos island, noticed the arrival of a male cactus finch, a non-native species. This male mated with a female ground finch and the two produced fertile offspring. Almost four decades later, the approximately thirty individuals who are the progeny of that original pairing are still being observed. "This new finch population is sufficiently different in form and habits to the native birds, as to be marked as a new species . . ." The various finch populations on the island do not, as a general rule, interbreed. As the fertile offspring from that first cross-species mating between cactus and ground finch makes clear, there are no genetic barriers to reproduction. The constraints are behavioral. How are we to explain why that uncharacteristic initial mating took place? Who can say what those two birds might have been thinking? Researchers speculate that the male flew in from an island 65 miles away. Why would he undertake the challenges involved in making a return trip of that length when there was one pretty little medium ground finch willing to grant him a sweetheart, perhaps not in every port, but certainly in this one? And why would she choose him? I suggest that his "outsider's appeal" was probably sufficiently strong to overcome any doubts that may have lingered in her heart or at least in the left hemisphere of her brain.

Researchers point out that the new finches on Daphne Major are larger than the resident birds, a fact which has enabled them to occupy a previously unexploited niche. Once the chicks from this first cross-species mating had hatched and fledged, natural selection determined their fate. There was an uncontested source of food, and so they survived. Dubbed the "Big Bird" population, this new species has been tested and confirmed to be genetically distinct from the other finches on the island. Other than that first pairing, why wouldn't the new species breed with the old and have the resulting hybrids get absorbed into the existing population? Lacking a risk-taking female drawn to a dramatically different stranger, what keeps the species separate? A behavioral barrier. Native finches seem unable to recognize the songs of the new males and won't mate with them. There we have it: sexual selection determining who gets born, natural selection determining what happens once they get busy figuring out how to survive. If there is nothing for the new sexually-selected species to eat, they'll die out. But if, as in this case, there is an available niche in the form of a seed that Big Birds are better able to harvest than the smaller native species, they will thrive. It's an example of natural selection and sexual selection working together to rapidly produce a new species. Professor Roger Butlin, a speciation expert who wasn't involved in the study, concludes, "Evolution, in general, can happen very quickly ... If you just wait for mutations causing one change at a time, then it would make it more difficult to raise a new species that way. But hybridization may be more effective than mutation."

I would agree. Hybridization happens when individuals from two species thought to be distinct from each other mate and produce viable offspring. Simply put, hybridization relies on sexual selection. And since evolutionary biologists agree that female choice guides reproductive decisions, perhaps it's time to pay closer attention to just what it is that females are doing. There is no better place to start than with *The Descent of Man and Selection in Relation to Sex.*

Chapter 4: Female Choice and the Origin of Man

There are a couple of things that really puzzle me. After all the years I've spent reading, thinking, and writing about evolution, especially about the weighted importance of natural versus sexual selection, how is it that researchers—both men and women—are so willing to agree that it's those sporting the ovaries who are in charge of reproductive decision-making? After all, it's not as if we've been particularly willing to give females credit for their achievements in any other field of endeavor. In *The Descent*, Darwin noted that advanced intelligence is what sets humans apart from other species. Since he argued that sexual selection driven by female choice was the force that shaped human evolution, and since women selected for intelligent men, he concluded—with a twist of logic that is best accounted for by his belief that sexual selection resulted in sex-related traits—that the average mental ability of men can be expected to be above that of women. In subsequent pages, Darwin expanded on his previous point by explaining how it is that women managed to end up with much intelligence at all: "It is indeed fortunate that the equal transmission of characters prevails with mammals, otherwise it is probable that man would have become as superior in mental endowment to woman, as the peacock is in its ornamental plumage to the peahen."

Darwin, who was at times willing to speculate on the inheritance of acquired characteristics as an evolutionary force, discussed just what might be involved in raising female intelligence closer to the level he attributed to men. "In order that woman should reach the same standard as man, she ought, when nearly adult, be trained to energy and perseverance and to have her reason and imagination exercised to the highest point; and then she would probably transmit these qualities chiefly to her adult daughters. All women, however, could not be thus raised, unless during many generations those who excelled in the above robust virtues were married and produced offspring in larger numbers than other women." Wow. Where to even begin unpacking this bit of muddled analysis? It's obvious that Darwin thought women—or at least women of a certain class and ethnicity—ought to have lots of babies. Since he subsequently revealed himself to be far less enthusiastic about *Irish* women's demonstrated skill with procreation, he must have been talking in the above paragraph about *English* women. Darwin's class biases and failure to understand the ways in which human economic behavior differed from the economic behavior of other species trapped him in what I refer to as a "Malthusian fallacy". Unfortunately, this fallacy and its emphasis on sheer numbers as an accurate measure of biological fitness continues to infect our conventional-wisdom understanding of how natural selection works, but more of all that later. For now, I'd like to point out that there is at least one practical barrier to raising female intelligence via the method suggested by Darwin. As recent demographic research makes clear, the most effective way to *reduce* birth rates is to educate females. All that training "to energy and perseverance" would have had the exact opposite effect of what Darwin hoped to achieve. Just what *is* the level of intelligence required in order to proficiently exercise female choice? Whatever amount it is, Darwin did not believe that women of his time had enough to competently execute the task. Although Darwin was very clear that it was the female of the species whose taste helped modify the traits subject to sexual selection, he was unwilling to grant his female contemporaries this power. He concluded that although female choice had undoubtedly been an evolutionary force in our distant forbears or even among "primitive" tribes, it no longer operated in "civilized" society.

Before deciding to marry his cousin (an example of "linebreeding" that was not unusual in his day and time), Darwin constructed lists both pro and con in order to help him determine whether or not he ought to settle down. The entries lead me to conclude that he assumed the choice was all his. Who knows what the deciding factor might have been? He compared the disadvantage of time lost to work to the advantage of enjoying music and "female chit chat." His expressed concern about a possible reduction in means does not appear to have been either well-founded or borne out, at least not in his eventual choice of a wife. Emma Darwin, nee Wedgwood, was a potential heir to a pottery fortune. In fact, Charles and Emma were *both* Wedgwood descendants, the lucky grandchildren of Josiah and Sarah. Linebreeding in humans may be better at consolidating wealth and controlling resources than it is at guaranteeing genetic superiority. That said, I don't mean to imply that the childhood deaths of three of Darwin's ten children is an indication of genetic weakness brought on by inbreeding, but think it is, instead, a case more easily explained by childhood mortality rates in the 1800s. Despite his expressed concerns about marriage interfering with his work, Darwin, a member of the landed gentry who was not forced to rely on formal employment, found time to study and write both before and during his marriage. After pages of pro-and-con, back-and-forth dithering, what finally tipped the scale in favor of matrimony? Was it simply because he saw the companionship of a wife as being in some way "better than a dog anyhow" or because, when all was said and done, "There is many a happy slave"? How's that for a shining example of sound reproductive decision-making in *"civilized"* times? Thank goodness that, despite Darwin's doubts, sexual selection via female choice was almost certainly operational in 1838 and continues to remain operational today. Unwilling to assume that Emma was simply a passive bystander in the courtship game, I would love to have had the opportunity to interview her. I am quite willing to bet that she had targeted her cousin Charles as a man worthy of her consideration and had identified him as a potential partner long before he came to view her in the same light. After all, it's not as if Emma had simply been sitting around waiting for someone to decide to marry her. She had turned down at least two marriage proposals while the young naturalist was off exploring distant ports and recording the insights on biological variation made possible by his journey on the H.M.S. Beagle. Once Darwin returned to England, I'm curious about the tactics Emma may have employed, consciously or not, to get herself into his field of vision, to bring herself to his attention. Through what signals did she communicate that she now regarded him as something other than cousin? Through what lens did she evaluate his potential as a husband and father? *Darwin* may not have been looking for anything more in the way of fellowship than what is typical of the bonds we have with our canine companions, but had that been a deciding factor for *Emma*, I suspect she simply would have gotten herself a Cavalier King Charles Spaniel. From all reports, once married, the Darwins had a loving, devoted, and mutually supportive relationship, and for that fact, I'm willing to give much of the credit to Emma.

Given the scant faith that Darwin, generally considered to be the father of evolutionary theory, had in the intelligence of females, how is it that he and those who followed him were able to reach a consensus on the ubiquity of female choice? In *The Descent of Man and Selection in Relation to Sex*, he cited example after example of female choice operating across a wide range of species, drawing not just on his own observations but on those of dozens of other naturalists as well. Perhaps the reason he and others working in this field so readily agreed on the power of female choice was that they didn't—and still don't—see sexual selection as concerned with making choices much more substantive than those relating to secondary sex characteristics or beauty. Despite Mr. Darwin's belief that female choice was no longer operational in humans, if the 66 women I interviewed are to be trusted—and I think they are—intelligence ranks just behind economic clout as one of two traits most predictive of a man's sex appeal and reproductive success. A woman was more likely to have children with a man she described as intelligent than she was with one who failed to pass her personal IQ test. Intelligence is a trait that continues to influence reproductive outcomes. In the study I conducted, the determination was a subjective one, largely based on a woman's assessment of a man's intelligence relative to her own. Darwin was correct in believing that women are drawn to smart men. That almost certainly was one of the traits that endeared him to Emma. However, let's not give the human brain more credit than it is due. Evolution's first concerns are reproduction and survival. Rational thinking is selected for only as long as it doesn't interfere with the push toward procreation. Love may not be entirely blind, but it is rather near-sighted, especially when it comes to a man's ability to see himself as the target of a woman's choice. There is wisdom to be found in old adages. "A man chases a woman until she catches him." Not even Darwin was able to completely distance himself from a million years of evolutionary programming.

Other than granting that sexual selection can explain the evolution of beauty, the only other time some biologists turn to it is when a trait appears to be *maladaptive,* as in the case of the outsized antlers of the Irish Elk and the elongated tail feathers of the widowbird. Darwin did, at times, allow that sexual selection influenced traits other than beauty. However, once racial attributes and sex differences were accounted for, he seemed to think that natural selection was all an adapting species required. Although I'm not willing to assume, at least not without further study, that traits such as the ones described as maladaptive are detrimental to an individual's reproductive success, that is not the same as saying that all attributes shaped by female choice are beneficial. Some traits that prove advantageous in the early stages of speciation might be less adaptive in a mature species. This is especially true in the human case. Alfred Russel Wallace informed Darwin that early humans were able to survive only when we figured out how to shape the world to accommodate us rather than having to wait for the slow process of natural selection to fit us to the environment in which we found ourselves.

We've all been told that most mutations are deleterious, and in genetic disorders such as cystic fibrosis and sickle cell anemia, they obviously are. As previously mentioned, great apes have 24 chromosome pairs, humans have 23. Why should we assume that the rather sudden shift from a great ape's 24 chromosome pairs to our 23 occurred without a certain amount of potential trauma? We avoided elimination only by figuring out a way around natural selection's blind indifference to the fate of creatures born lacking biological fitness. The human approach to survival was and remains profoundly and radically different from that of other species. We became the actors, rather than the acted upon. Hold this distinction in mind. We will return to it and all its implications in later chapters. For now, simply note that a trait—the ability to change the natural world—that was *in the beginning* absolutely necessary to our survival may have become maladaptive in 2021. Our very skill at exploiting resources has altered the face of the earth and the atmosphere surrounding it. It is difficult to argue for the biological fitness of a species that destroys the environment necessary to it and the interconnected web of life synergistically linking all species on the planet. Sexual selection, with its dependence on subjective decision-making, is more vulnerable to errors in judgment than is the cruelly objective live-or-die imperative of natural selection. Female choice has almost certainly amplified some human traits that are no longer contributing to our survival as efficiently as they once did. However, I won't give natural selection exclusive bragging rights for *every* positive adaptive change that has made us who we are. Female choice has had at least something to do with a lot of our praise-worthy qualities. Language. Intelligence. Economic prowess. A technological talent for exploiting the resources necessary for life. Kindness. Devotion to kin. Generosity. Social skills. Cooperation. How did we get to be the species that has demonstrated unprecedented genius, a genius possessed of both an upside and a downside, for dominating life on Earth? Natural selection? Sexual selection? Both. The two forces enhance and complement each other.

Because I've been so frustrated by the failure of evolutionary biologists to grant sexual selection equal billing with natural selection, I plead guilty to being an overly zealous proponent of its role in human evolution, sometimes failing to consider the possibility and importance of other factors. As his preface to the second edition of *The Descent of Man and Selection in Relation to Sex* makes clear, Darwin shared my frustration—if not the zealousness of my advocacy. By the way, I'm charmed that in the Table of Contents of this same edition, Darwin gives his readers a choice of title: "The Descent or Origin of Man." Had he settled on "Origin" and avoided reminding his audience so blatantly that we were *descended* from apes, I wonder if Darwin might have found a way around the public's failure to accord sexual selection the respect that he felt it deserved.

> Again, it has been said by several critics that when I found that many details of structure in man could not be explained through natural selection, I invented sexual selection; I gave, however, a tolerably clear sketch of this principle in the first edition of the "Origin of Species" and I there stated that it was applicable to man. This subject of sexual selection has been treated at full length in the present work simply because an opportunity was here first afforded me. I have been struck by the likeness of many of the half-favorable criticisms on sexual selection, with those which appeared at first on natural selection, such as that it would explain some few details but certainly was not applicable to the extent to which I employed it. My conviction of the power of sexual selection remains unshaken, but it is probable, or almost certain, that several of my conclusions will hereafter be found erroneous; this can hardly fail to be the case in the first treatment of a subject. When naturalists have become familiar with the idea of sexual selection, it will, I believe, be

> much more largely accepted, and it has already been fully and fairly received by several capable judges.

There are still evolutionary mysteries to be unraveled. As the numerous examples of convergent evolution described by modern biologists make clear, factors that we do not yet fully comprehend are almost certainly at work. The more we learn about genetic complexity and the interactions between RNA, DNA, and the mostly unexplored recesses of dark genetic material, the more we realize how much is still gloriously unknown. Darwin hints at the possible evolutionary impact of some of these in the sentences that precede the above quote. He speaks to "the inherited effects of use and disuse," "the direct and prolonged action of changed conditions of life," the "occasional reversions of structure," and "correlated growth" such as that seen when ". . . various parts of the organism are in some unknown manner so connected, that when one part varies in structure so do others; and if variations in the one are accumulated by selection other parts will be modified." No one was a more astute observer of the various structures found in plants and animals than Darwin. He made note of many things that didn't seem to fall within the range of explanations that could be attributed to either natural or sexual selection.

As has often been pointed out, Darwin was unaware of the genetic correlates of the morphological and behavioral variations he documented. Once genes were identified as the basis of inheritance, it was easy to assume a one-to-one relationship between a gene and the trait under observation. But, as we now know, it's more wonderfully complex than that. For example, we are just beginning to grasp the intricate microbiology orchestrating what we refer to as epigenetic inheritance. Those who claim support for the inheritance of acquired characteristics by pointing to well-publicized examples of the metabolic changes identified in the descendants of victims of starvation are probably jumping the gun. Epigenetic inheritance and the inheritance of acquired characteristics are not necessarily the same thing. The cases most often cited involve the children of Holocaust survivors and of those who were subjected to the Dutch Hunger Winter near the end of World War II. The progeny of both groups appear to be at greater risk of obesity than would be predicted by their eating habits alone. It has been *suggested* that these epigenetic changes prepare the descendants of the food-deprived for an environment in which there is as little access to food as there had been in concentration camps or among those subjected to the deliberate Nazi blockade of the Netherlands. As told in the popular press, the experience of caloric limitation in the parental generation somehow gets encoded into the genetic blueprint concerned with nutrient absorption in their descendants. Even when subsequent generations were reared in conditions of relative plenty, they appeared to metabolize food as if their bodies had somehow been prepared to extract every single calorie with extreme efficiency. It's as if the children of women who were pregnant under those brutal conditions had been pre-adapted to scarcity and were subsequently able to maintain or gain weight on far less food than was typical of their peers whose ancestors had not been subject to extreme deprivation. None of this is a straightforward as the headlines would have us believe. In fact, the children of Holocaust survivors have a stress hormone profile that itself might be more responsible for insulin resistance and obesity than is any supposed metabolic "efficiency."

Be that as it may, there is enough evidence of epigenetic inheritance in both plants and animals to have inspired some first-rate research. I am fortunate to live a short distance from a number of major universities, enabling me to take advantage of the many lectures that these institutions make available to the general public. I was especially impressed when Karin Michels, Sc.D., a 2017 Radcliffe Fellow, gave a presentation challenging the permanence of epigenetic inheritance. Michels has the rare ability to address a lay audience and describe, in riveting fashion, something as complex as the molecular interactions involved in DNA methylations influencing gene activation. She gave a precise account of just how the expression of genes is turned on and off. Those of us fortunate enough to be in the lecture hall leaned forward in our seats and listened as intently as if we were keeping track of the plot in an exciting who-done-it mystery. I was not alone in my enthusiasm. Despite my lack of any formal background in the field, I came away from her talk convinced that the claims of permanent genetic changes resulting from environmental events, such as starvation, are as yet unsubstantiated. In order to make the case for *permanent* genetic changes, they would have to be shown to persist for more than four generations when the mother experiencing the event was pregnant with a female fetus at the time of the precipitating incident, or more than three if she was pregnant with a male fetus. The reason for the sex-related disparity in the number of generations is that a female is born with ovaries containing all the eggs that will develop over her lifetime. Any environmental impact on a woman's genetic material will likewise impact all the eggs of any unborn daughter she might happen to be carrying. Males don't begin producing sperm until they reach sexual maturity. My layperson's conclusion? Epigenetic change happens but is generationally limited and does not offer proof of the inheritance of acquired characteristics.

One of the more benign criticisms that were leveled at sociobiology, a field of study undertaken by those of us who wanted to explore the possible inheritance of traits impacting both biology and *behavior*, was that it was riddled with "just-so" stories. The term sociobiology was itself popularized when E. O. Wilson used it as the title of the book he wrote to introduce readers to ongoing research documenting the genetic logic behind the endeavor. I suspect that he and others involved in this developing field were caught off guard by the viciousness of the attacks that followed the publication of his book. The most outrageous of these charged that we were eugenicists at heart and that our research would encourage a return to the horrors of racial and ethnic stereotyping and the murderous "final solutions" associated with those dark chapters in human history. In fact, we were simply looking to understand the origins of human behavior. If anything, we were documenting physical and behavioral universals, the existence of species-specific traits that defined our common humanity. There was no covert advocacy for the role of nature over nurture. We accepted that human behavior was shaped by both and pointed out that there was much about the influence of genes on behavior that was not yet understood. In any event, much of the controversy faded away once the human genome was sequenced. I hope sociology classes no longer begin with the statement, "All human behavior is learned," that I encountered both as an undergraduate and when later pursuing a master's. The term sociobiology remains toxic to many. The preferred nomenclature is now evolutionary psychology. I stubbornly stick with the original designation, liking its inclusive reference to the evolution of both the social and the biological.

Although I don't advocate for a return to the days when the equivalent of nursery tale "just-so" stories were used to explain evolutionary outcomes, I confess to a personal affection for "just-suppose" or "what-if" speculations. I sometimes indulge in these sorts of conjectures when I find accepted theory—such as Darwin's heavy reliance on a slow, numbers-dependent evolutionary process driven almost exclusively by natural selection—inadequate or misleading. And I sometimes seize on the unexplored implications of solid empirical research to entertain myself in a playful "what-if" search for answers to questions I hadn't realized I had. Indulge me for a moment while I speculate on how epigenetic evolution, as illuminated by Karin Michels, might possibly facilitate permanent changes in an organism whose genetic expression has been temporarily altered by abrupt environmental changes. What might be the purpose, or functionality, of an epigenetic change that disappeared after only a few generations? On the surface, these generationally limited adaptations would appear to be a kind of quick-fix response to dramatic environmental changes that were themselves not likely to last. But what if the change was more or less here to stay? Suppose, for example, that the precipitating factor was a famine that turned out to persist for more than a few seasons. Imagine that it was caused by the not entirely unlikely scenario of climate events that rendered wide swaths of the world unable to produce enough food to nourish the population. Epigenetic changes in the fetuses of women pregnant during the initial season of famine would enable their kids to survive. Now should climate change magically reverse itself, those children would enjoy no particular metabolic advantage, and over the course of three or four generations, the temporarily adaptive advantage of the epigenetic change would disappear. Given a human generational mean of about 25 years, the descendants of mothers who had been pregnant when the famine first hit would have a survival advantage that might carry a small group of the population through almost 100 years of food scarcity. During that time, women who became pregnant would begin the epigenetic cycle anew. Is there a way to make survival mechanisms that were initially introduced via a bit of epigenetic magic a *permanent* feature of all the individuals in a population? A hundred years gives any species with a 25-year generational span a lot of time for both natural and sexual selection to sift through the potential adaptive value of existing genes that influence metabolic rates. If the dramatic change in an ecosystem's ability to produce abundant crops failed to automatically correct itself within a century, the epigenetic alterations would enable at least some individuals in the population to survive and allow ample time to apply selective pressure to a whole range of genes impacting metabolic efficiency. Those who didn't die of starvation would be both available and attractive to potential reproductive partners. Women select men based on their ability to be good providers. If the male *Anolis garmani* lizards studied by Robert Trivers can advertise their skill at resource accrual by displaying enhanced body size, I think it's safe to assume that human males could do the same. In an ecosystem where most people were emaciated and starving, I think it would only make sense for a woman to fall for a man with just a bit of weight on him. Once again, I am willing to grant human females an intelligence equal to that found in reptilian brains. The examples I've outlined in my early chapters indicate that sexual selection can produce evolutionary change in only a few generations, enough to make epigenetic changes permanent should the conditions that first inspired them to persist. The selection would not necessarily be for the *epigenetic* change that had enabled temporary survival but for the existence of any small genetic variation that would give the individual carrying it an expanded window of time in which to employ the more traditional forces of evolutionary adaptation. Temporary epigenetic changes give some individuals a profile that enables them to survive tough conditions while also providing an opportunity to see whether or not the altered condition is going to be the "new normal". If it appears that the climate changes responsible for famine conditions are permanent, there is a window of time in which individuals can select for traits that will spread through a population—particularly if it's a small, isolated population—and become a permanent part of a species' genetic arsenal.

Chapter 5: The Differences that Make a Difference

It is generally accepted that our species evolved very rapidly. The earliest known fossils with features that resemble anatomically modern humans are about 300,000 years old. That's a drop in the bucket of evolutionary time. In an earlier chapter, I alluded to my belief that the original speciation event was probably fairly abrupt. Researchers trying to understand anomalies in brain size linked to certain genetic disorders have recently identified genes impacting the volume of the brain. Christopher Walsh, a neurobiologist with expertise in genetic analysis, suggests that an increase in brain size that occurred about 200,000 years ago could be related to the fact that humans have duplicate copies of what is designated the Notch 2 gene, while the great apes have but a single copy. If he is correct in the timing he suggests, this mutation on chromosome 1 would have occurred around 100,000 years *after* the hominids responsible for those 300,000-year-old fossils, the fossils that already had brains resembling ours. There is more to being human than can be told by the simple weight and measure of our brains. I think it's possible that a lot of the changes that define us spring, not from a mutation in a single gene, but from a far more significant shift in genetic architecture, one that reduced the number of the chromosomes themselves and had system-wide ramifications. Having long been drawn to Stephen Jay Gould's and Niles Eldredge's theory of punctuated equilibria and their observation that the origin of species was probably not as gradual as Darwin had led us to believe, I often toyed with the idea that incidents of genetic nondisjunction could have accounted for the sudden appearance of new species. However, it was not until relatively recently that I became aware of the role *end-to-end* chromosomal reduction might play in speciation events.

A study conducted in 1982 by Jorge Yunis and Om Prakash at the University of Minnesota documented a shift in genetic architecture that could have been the source of an evolutionary event that would fit nicely within the Gould/Eldredge model. Operating on the evidence that chimpanzees, gorillas, and orangutans are our closest relatives, they took photographs of each set of the human and great ape chromosomes and then lined them up, hoping that such a comparison would provide clues to the differences that might help explain why we're so filled with fancy talk and why orangutans, gorillas, and chimps enjoy hanging out in trees. The casual remark about talk and trees is all mine.

Yunis and Prakash stuck to a more rigorous research standard. After careful study, Yunis and Prakash concluded that 18 of the 23 chromosome pairs were almost identical. In some other locations, they did see evidence of occasional inversions—flips that reverse the orientation of the genes in a small segment of a chromosome—that occur even in individuals within an existing species. Most of these inversions don't make a noticeable difference in expressed morphology or behavior when they show up in either human or non-human primate chromosomes.

In humans, we know that there are segments of some chromosomes, perhaps due to an inherent weakness in whatever "glue" contributes to the overall cohesion of that particular block of genetic material, that are more vulnerable to inversion "errors" than are found in other segments of the same chromosome or in various places in altogether different chromosomes. It is not usually a big deal. But there is one difference between the genetic architecture of humans and that of the great apes that *is* a very big deal: a disparity in the number of chromosome pairs. This provides an obvious starting place when trying to figure out how the reduction happened and what its role might have been in directing the development of a new branch on the family tree. Think about the overall genetic impact of such a dramatic change. Could we have lost an entire chromosome and all the genetic material it contained and still have managed to survive? Not likely. A far more persuasive explanation is that two chromosomes somehow fused into one. According to the 1982 study, the most probable candidate in support of the fusion theory was to be found in the structure of human chromosome number 2. The two separate *halves* of human chromosome 2 corresponded to two separate *chromosomes,* 2A and 2B, found in great apes. These similarities offered evidence of an end-to-end fusion. In 1991, a follow-up study conducted at Yale took advantage of progress in the sequencing of the human genome and a near-decade's worth of related technological advances to add additional data in support of the original conclusion. I think it likely that this fusion of great ape chromosomes 2A and 2B into the single chromosome 2 found in humans provided a single dramatic speciation event from which all other differences between ourselves and our closest primate relatives followed. It involved so much more than a small mutation in a single gene and almost certainly resulted in the differential expression of genes in many seemingly unrelated parts throughout our bodies. Could we have gone almost overnight from hairy apes to naked apes? By looking at some of the known outcomes of chromosomal mutations *within* our species, we can gain insight into the sorts of changes in genetic expression that might be expected from something as dramatic as a change in the number of chromosomes defining the boundaries *between* species.

Anyone familiar with high-risk pregnancies, fertility clinics, embryo implants, miscarriage, and all the associated heartbreak quickly learns that not all chromosomes are created equal. There is a whole range of genetic syndromes associated with the kinds of structural damage that occur when chromosomes fail to completely and cleanly separate and duplicate into copies with all the necessary alleles. These various forms of atypical sticking together and doubling up are loosely referred to as nondisjunction, a failure of homologous or sister chromatids to separate properly during cell division. They can occur in various stages of either meiosis or mitosis. Sometimes a small chunk from one chromosome attaches itself to a neighbor, forming a mosaic. Sometimes an extra copy of a chromosome results in an individual possessing three versions of a particular chromosome where we would have expected to find just two. Trisomy 21 occurs when there are three copies of chromosome 21, and babies born with this extra copy are diagnosed with Down syndrome. Some chromosomes appear to be more prone to trisomies than others. Even within chromosomes, some segments are especially vulnerable to error. As I've said, not all chromosomes are created equal. Some anomalies occur with enough frequency to have been dubbed with identifying nomenclature, such as Edwards syndrome (trisomy 18) or Patau syndrome (trisomy 13). Most trisomies are not compatible with live birth, and, even in those that are, it's difficult to predict in advance the *extent* of an expected change in genetic expression, such as body configuration and intelligence, but there is general agreement on *where* those changes will express themselves. Height is an easy one to document, as is the kind of academic proficiency measured by standardized IQ tests. Others are more surprising. Children with Prader-Willi syndrome tend to have a stimulus-response reaction to food, have difficulty regulating their caloric intact, and will struggle with weight gain and its related medical issues throughout their lives. In other cases of nondisjunction, whole chunks of chromosomes are simply lost. Differences in our male-female sex chromosomes may have had their origin in the loss of a segment of the X chromosome. An X with a missing arm becomes a Y. Just look at the range of body-wide gene expressions associated with that simple change. Trisomic-style variations in the standard XX female, XY male model are among the more common and survivable structural chromosomal anomalies. They occur with enough frequency to have names. Turner syndrome. Kleinfelter syndrome. The examples listed above are but a few of the identified body-wide changes in gene expression that result from a simple rearrangement in chromosomal architecture. We think of all mutations of this sort as deleterious, and certainly the association of many of them with miscarriage reinforces that view. However, as the generally sunny dispositions of children with trisomy 21 would indicate, differences can skew toward the positive. In addition, some mosaics or inversions are so infrequent as to have not been labeled as identifiable syndromes. The only time they come to anyone's attention is if they result in an effect that is somehow problematic. If a tiny mutation doesn't cause problems, it is not often revealed by a follow-up examination of a child's genome. Perhaps we need to give further thought to the view that most mutations are deleterious. It may simply be that we are more likely to *notice* the ones that confer a trait that results in developmental challenges. We are far less likely to look for genetic mutations as a way of explaining a child with an unusual talent or other sort of advantage.

Chromosomal changes of the sort hypothesized in the end-to-end fusion of great ape 2A and 2B into human chromosome 2 are not as common as the sorts of nondisjunctions that occur during typical human cell division. Individual chromosomes are protected by telomeres that protect against just such end-to-end fusions and work to maintain the kind of chromosomal integrity that serves as a physical reproductive barrier between species. Nonetheless, these fusions do happen. A number of closely related species—domestic horses, wild Asian horses, donkeys, many species of mice, etc.—almost certainly acquired the difference in chromosomal count that prevents them from breeding with closely related species through fusions of the sort described above. We've long been told that individuals with different numbers of chromosomes either can't produce viable offspring or can produce only those that are themselves sterile, the most famous example being a cross between a horse and a donkey resulting in an infertile mule. If a fusion of chromosomes 2A and 2B provided our primary speciation event, why didn't that innovative line simply die out without ever reproducing? As I have noted, not all chromosomes are created equal. This particular chromosome may be more vulnerable to end-to-end fusion than others in our repertoire. In addition, at the time of the event, there could have been some environmental incident that increased the odds of mutation, thereby providing similar individuals with whom the new-model primate could reproduce.

Several decades ago, both parents in an otherwise healthy Spanish family each exhibited a fusion of chromosomes 13 and 14. Three of their six children were born with 44—rather than the standard 46—human chromosomes; they went from 23 chromosome pairs to 22. I find this example fascinating for a couple of reasons. It clearly demonstrates that chromosomal reduction is not always fatal or even, at least in this case, particularly deleterious, but of special interest to me is that *both* parents had a single fusion of 13 and 14. Remember my earlier observation that individuals in all species seem to have an almost subliminal ability to recognize other individuals with whom they have shared genes? It's a possible explanation for why married people have a greater number of genes in common than would be predicted by chance alone. Might this gene-recognition-theory-of-attraction explain how the Spanish couple with the fusion of 13 and 14 ended up marrying? Or were they simply from a small community in which there was already a high degree of relatedness between individuals and one in which there were relatively few potential partners? What attracted these two parents to each other? Although it's possible to do a Google search and view images of the chromosomal architecture involved, I can find no biographical information about the couple themselves. I would not expect the geneticists conducting the sophisticated chromosomal analysis to have allowed themselves to be distracted by shifting their focus to courtship behavior. That's my piece of the puzzle, and each additional bit of data that they gather happily adds to the fuel that fires my imagination.

Suppose that the end-to-end fusion that combined great ape chromosomes 2A and 2B into human chromosome 2 really was the speciation event that set us apart from our closest primate relatives. Would there have been obvious physical changes? It doesn't sound as if the fusion of 13 and 14 resulted in anything particularly dramatic in those carrying it, individuals who were described as "otherwise healthy." But if the formation of a single chromosome 2 was as big a deal as I think it could have been, where should we look for significant traits associated with the change? A good starting place is to consider the obvious physical differences between us and the great apes. Large cranial capacity. Upright posture. General loss of body hair accompanied by an abundance of hair on the head and around the pubic area. Those are the differences we might note at a glance. And, of course, this is all speculation. Those obvious differences could well be the result of a couple of million years of natural and sexual selection acting on variations made possible by the initial reduction in the number of our chromosome pairs. Be that as it may, I would be willing to bet that there were some fairly immediate and obvious changes in the physical appearance of a baby born as the result of the chromosomal fusion that set us on the path to *Homo sapiens*. Although the *sapiens*-like shift in brain function that we humans like to claim as evidence of our superiority would not have been obvious at birth and any behavioral and reproductive patterns made possible by changes in the brain would emerge only as the child matured and acquired skills, there is at least some evidence that makes the fusion of these particular chromosomes a likely candidate for an altered expression of genes relating to our brains. Manuel Ruiz Rejón, a professor at the University of Granada, briefly summarized the potential tissues and organs that would have been impacted by a fusion in this region. According to Rejón, there are several genes in the region of the fusion that "are expressed more intensely in our species than in that of the large apes. These genes are expressed, above all, in highly significant tissues and organs, such as the brain and the gonads." Since most researchers concerned with identifying the traits that set humans apart begin by looking at the brain, this is indeed a tantalizing clue in support of chromosome 2 as a central place to start looking for the key to human origins. Given my focus on sexual selection, I was even more taken with his mention of the significance of genes regulating gonadal tissue, in other words, of genes regulating the testes and ovaries. There are dramatic differences between the reproductive behaviors of humans and chimps. The sixty-six women I interviewed spent between two and five hours sharing the secrets of their sexual strategies and choices with a tape-recorder and me. Any apparent strategizing by a female chimp appears to be fairly hormonally driven and limited to the ovulatory stage of her reproductive cycle. Even if a chimpanzee could talk, I don't think it would take her five hours to explain how she had zeroed in on the male that impregnated her. In fact, there doesn't appear to be much zeroing in at all. A female chimp lets her body do the talking. Unlike human females, she doesn't waste time with flirtatious eye contact across a crowded room or with rearranging a morning commute so that she now takes the same train as the man who caught her attention earlier in the week when car trouble forced her to resort to public transit. When a female chimp is in heat, the skin around her genitals becomes pink and swollen. That and the accompanying pheromonal signals send a pretty unambiguous and indiscriminate advertisement out to all the males within the community. And just in case one male doesn't get the message and believe me, almost all of them do, she is not above thrusting her inflamed hindquarters right up into his face.

Many years ago, back in the mid-seventies, when I was a graduate student just starting my doctoral research, I was trying to help one of my professors understand why it was important to consider the role that sexual selection had played in driving human evolution. I reminded him of Darwin's belief that sexual selection operated through female choice and went on to explain why I thought that by interviewing modern women, I might get some insight into how the traits that influenced the choices females made today could hint at the selective processes that had shaped the human animal historically. This professor, a primatologist, shook his head, paused for a split second to stare at me in stunned silence, and then burst out laughing. "No one who has ever witnessed the wild orgy that results when a female chimp is in heat—it's complete pandemonium—could possibly argue that there is any room for either sexual selection or choice in our evolutionary history. So much for your theory about the importance of sexual selection! Females exercise no choice at all!" As I watched him walk down the hall still chuckling to himself, it dawned on me that the difference that distinguishes chimps from humans just might have had its beginnings in concealed ovulation and its equally innovative sister, continuous sexual receptivity. Even then, I was very much aware that human females were/are unique among mammals in general and primates in particular in our failure to blatantly flaunt the fact that we have eggs just waiting to be fertilized, but not until that professor laughed at my research proposal did it dawn on me that a *single difference* in mating behavior could hold the key to explaining how all the other differences between us and the great apes evolved. His dismissive amusement fueled my eureka moment. Human females are sexually receptive throughout their reproductive cycles. Female chimps are interested in sex only in the fairly narrow timeframe when they are also fertile. Human females don't advertise their fertility, and the potent combination of concealed ovulation and a healthy sexual appetite give human females a tremendous amount of authority in shaping our own lives and those of our children. It should be apparent by now that I think we have overlooked the agency of females in all species, but nowhere have we been as blind to its power as we have been in the case of humans. Women are very skilled at playing our cards close to the vest. Reproductively, we've employed this cloak of invisibility and used it to our advantage. It's our reproductive superpower, one that gives us time to learn more about that stranger on the train, to appraise him on a variety of traits—resource accrual, intelligence, social skills, willingness to commit, generosity, humor, and a way with words. Because, here's the thing: that big brain of ours does more than complicate childbirth. It also enables a subtle rewiring that facilitates language, the tool that provides a unique adaptive advantage and shapes the cultures and technologies that make it possible for this otherwise poorly equipped naked ape to survive.

As mentioned earlier, I've only recently become aware of research substantiating the fusion that chromosomally separates us from the great apes. In all the years since my grad student days at Rutgers, I've been a somewhat kooky voice muttering about the failure of neo-Darwinists to give sexual selection its evolutionary due. I've been known to pigeonhole speakers at conferences in an effort to get them interested in using their youth and sophisticated understanding of genetic analysis to hunt down the gene or genes that might be associated with concealed ovulation in humans. I have stooped to becoming the email equivalent of a bus station pest handing out religious pamphlets in the Port Authority Terminal building in NYC. I've tried my darnedest to get someone, *anyone,* intrigued by the possibility that once the genetic marker for concealed ovulation was identified, we could then look at fossil remains and see if the existence of concealed ovulation correlated with the symbolic art and deliberately crafted stone tools that signal the arrival of the human animal. I am convinced that the evolutionary moon shot that enabled us to go where other species have failed to follow has its roots in reproductive changes that dramatically increase the strategic agency employed by human females. I admit to being a nuisance on the subject. It's one of the few things that makes me wish I were young enough and technologically skilled enough to tackle the research myself. I can understand, really and truly understand, why no one wants to listen to my rantings. I know what it's like when someone is urging you to do something that they can't or don't want to do themselves. When I meet a stranger at a party and identify myself as a writer, it's not uncommon for that person to start telling me about how I need to write a book detailing the story of the time when her grandmother's sister was kidnapped as a baby and the numerous and complicated ways that single incident has impacted all the lives in subsequent generations. My new acquaintance is perfectly willing to *give me the idea* if only I will agree to write the novel. I acknowledge that it *is* a fascinating story and one that would make a great book, but I have my own ideas about what might make an interesting topic. I don't want to tell someone else's story. I want to tell my own, but sometimes I need to borrow the words and grammar provided by the diligent research of others. I'm aware that, in doing so, I can create scenarios that take more speculative risks than these appropriately cautious scientists would allow themselves. I'm grateful that by pursuing their own passions, they have inadvertently provided me with directional signs that guide me in my personal quest to consider byways in the evolutionary process that have not been explored to my satisfaction. I'm admittedly not a specialist, not a technician. I'm a generalist, a daydreamer. When I see someone else diligently chipping away and creating exquisite stone tools, I immediately find myself constructing the story I'll get to tell those who later gather around the campfire. Only I won't limit myself to describing the tools or the process by which they were crafted. I'll see it all as being about so much more than brilliantly shaping stone. I'll want to enlarge the story, to weave in conjectures considering the ways in which those tools were used and how they might have helped determine the course of all subsequent human history.

Once I'd fixated on the unique nature of human reproductive behavior, I looked for research that had the stated intent of identifying regions in the human genetic sequence associated with concealed ovulation. Unable to find studies with that specific focus, I was ecstatic when I accidentally stumbled across Professor Manuel Ruiz Rejón's description of the unique characteristic of some regions of human chromosome 2, the one that I think made all things *Homo sapiens* possible. Although I've used this same quote just a few pages back, I repeat it here in order to reinforce the importance I ascribe to his observation. [On chromosome 2, there are genes that] "are expressed more intensely in our species than in that of the large apes. These genes are expressed, above all, in highly significant tissues and organs, such as the brain and the gonads." *The brain and the gonads*! The exact two places where I would expect to find consequential differences between humans and great apes! I've never met Dr. Rejón. But with these few words, he has given me hope that I might be on the right track, that the fusion of *two* great ape chromosomes to create *one* human chromosome might provide just the breadcrumbs I need to follow in order to fashion a story that could build on and complement the one Darwin gave us. In this chapter, I've hinted at the importance of concealed ovulation and its partner in power, continuous sexual receptivity. In later chapters, I'll remind readers of the special efficacy the combination of these two traits gives women in orchestrating our evolutionary journey. But any argument that credits ovaries with being the source of human exceptionalism meets with a lot more pushback than one that credits the brain. Therefore, I want to start by considering the human brain. In a way, I'm taking the path that offers the least resistance. There aren't many people who are going to challenge the notion that it's our brains that made us who we are, but, that said, we may be giving the brain a bit more credit than it necessarily deserves. At the very least, we may be misrepresenting just what it is that our brains have been shaped to do.

Chapter 6: My, What a Big Brain You Have!

Here's the thing to remember about the brain: it was not shaped by evolution to be a camera, tape recorder, or computer. It was shaped to enable us to survive and reproduce, and in order to be successful at those two tasks, it had to underperform in fairly significant ways when compared to each of the technologies listed above. The brain is not programmed to accurately record, store, replicate, and calculate all the data fed to us by our surroundings. For starters, it's really good at totally ignoring some information. There's the classic study in which participants are told they will be shown a short video of six people passing basketballs around. Three of the group members are wearing white shirts; three are wearing black. The subjects in the study are instructed to count the number of times those dressed in white pass the basketball to similarly garbed others, but they are not informed that while the ball is being tossed from player to player, a human dressed in a gorilla suit will wander into the middle of the action and proudly thump its chest. I've read many accounts describing this test of selective attention and knew that folks became so focused on counting the passes that half of them totally and literally ignored the elephant—or in this case—the *gorilla* in the room. Christopher Chabris and Daniel Simons, psychologists and co-authors of the book, *The Invisible Gorilla*, designed the experiment to reveal two things: that we miss a lot of what goes on around us and that we have no idea of just how much we are missing. The test is available online. Just Google "invisible gorilla." Even though I was aware that my results would be invalidated by the fact that I knew a gorilla would stand there pounding its chest as the balls whizzed by, I decided to take the test myself. Competitive by nature, I determinedly kept my eye on the ball and, less than a minute later had tabulated 15 passes. However, I had failed to see the gorilla. Since I'd known it was going to put in an appearance, I was certain that this first run-through must have been nothing more than a demonstration of how the actual test would work. Nope. It *was* the test. I may have been pretty good at tallying up the movement of the ball from one person to the next, but I had missed the very thing, the very *unusual* thing, that I knew in advance was going to be there. How was that even possible? So, I watched the video again and made no attempt to count passes this time. Sure enough, there was the gorilla, large as life. Even though I often talk with others about the selective nature of memory and perception, I was myself fooled. Taking the test was an instructive, if humbling, experience. So much for the brain's camera-like skill at videotaping. We are so bombarded with sensory input that, unless we learn to screen some of it out, we can find ourselves immobilized. There is something to be said for an ability to focus on the task at hand. Stopping to count the roses is not a great survival skill when one is being pursued by a lion or a gorilla.

Our memories are no better at accurately recalling events in our own lives than I was at noticing a gorilla on a basketball court. How many times, in the midst of a family argument, has one spouse stormed out of the room angrily shouting over their shoulder, "I wish I had a tape-recorder!" Although the presence of a tape-recorder might be able to resolve the conflict over who said what, it can't be guaranteed to save the marriage. However, what it might do is offer an instructive example every bit as humbling as my recent ball-and-gorilla-counting experience. The chances are good that *neither* party involved in a "tape-recorder fight" has accurately remembered what either one of them said or heard. Our brains don't operate like tape recorders. They help us survive by constructing narratives that enable us to get on with our lives. Our memories can be most *functional* when they work to create or maintain an image of ourselves that just may be *inaccurate*. Our brains tell us stories that serve to imbue our lives with meaning, that help us believe we are better, kinder, nicer, smarter, more honest people than we consistently manage to be. In a myriad of small ways, these self-deceptions provide the faith in our future potential that keeps us moving forward. If you want a really extreme example of the survival function of being able to put a positive spin on our behavior, consider this: individuals who have completely lost faith in their own value and purpose are vulnerable to suicide. In such a case, the ability to create fictions can literally keep them alive.

Many of the lectures that universities in the Boston area make available to the general public are in fields directly related to my research interests. Always eager to show off the charms of my home turf to visiting house guests, I was disappointed a few years ago when nothing relating to evolutionary biology was on the docket during the week a friend and colleague from California was visiting. So, in an attempt to demonstrate the range of offerings routinely available to me, I took her instead to lecture on a subject in which neither of us had a particularly compelling interest: memory. Elizabeth Loftus, a distinguished professor of social ecology and a lecturer of cognitive science and law at UC, Irvine, was speaking at the Radcliffe Institute that day. She not only convincingly described research that documented the ease with which false memories can be suggested into existence, but she also fooled the audience into creating a few flawed memories of our own. As soon as the applause following Loftus' presentation died down, Sarah and I turned to each other in awe. She found the words to accurately sum up the experience before I did, "Wow, that was life-altering." And it was, but not in ways that have always served me well, at least not when it comes to disputing events from our childhood with my two sisters.

When my younger sister, Gail, and I were little, I was entrusted to take her to a neighboring playground. Since Gail was not yet in school, I took my responsibility very seriously. I forced her to hold my hand all the way to the playground and wouldn't relinquish my grasp even when we got there. Terrified that she might get hurt, I wouldn't allow her to climb the monkey bars, go down the sliding board, or get on the old-fashioned swings. However, Gail had her own ideas about what a trip to the playground was supposed to involve, and determined to join the other children in all the fun, she finally broke free of my grasp and raced behind the busy swing set with me in hot pursuit. Up until this point in our shared memory of that day, Gail's recall and mine are a near-perfect fit. The promises we made to Mother. The weather. How carefully we crossed the street. Gail's rebellion against my authority. We agree on everything except the most important thing, the thing that made the day memorable—which one of us had her head whacked open by the heavy wooden seat of a flying swing. We both remember the quick sharp pain, the loud, single-note crack when it hit and we can both describe the immediate and frightening amount of blood. We agree that Gail sobbed uncontrollably, but that adds nothing in support of her version of the story. Gail was born with an abundance of empathy and cried whenever *I* got spanked, so, of course, she would have wailed when she saw me afflicted with an injury even more dramatic than being thrown across someone's knee. Gail/I can vividly remember the hot stickiness of the blood and how it pasted her/my white cotton blouse against her/my shoulder and chest. We are in complete accord that, because we were forbidden to get in cars with strangers, we declined the kind offer of a gentleman who wanted to give us a ride home. However, guilt over the anguish our refusal was causing this would-be rescuer propelled me to the window on the driver's side, where I accepted his neatly folded white cotton handkerchief to apply to Gail's/my bleeding head wound. I knew we weren't allowed to get close enough to any unknown vehicle that we could get pulled in, but he was clearly frantic with worry. I remember not wanting to hurt his feelings or cause him more grief. We'd been drilled about the distance we were supposed to keep between ourselves and men in automobiles, and Gail recalls panicking over my risk-taking. My sister and I are both certain that, with his car moving at a crawl, he followed us until he had seen us safely into our mother's arms. But we can't agree on whether it was Gail or me who pressed that beautifully laundered square to her head in order to avoid leaving a bright red trail splashed on the sidewalk behind us. Every detail of our story matches, except for which one of us ended up with stitches. Absolutely certain that my memory was infallible—I am, after all, seventeen months older than Gail—I have never been able to understand my sister's rigid unwillingness to admit that her account of things might be wrong. Well, Professor Loftus has ruined all that for me. I won't ever again be able to confidently win a family argument of this sort. Sarah was right. The Radcliffe lecture *was* life-altering. I just wish my sisters had also witnessed the presentation. My new understanding means that I will forever be the one who has to yield in these kinds of disputes. I've learned to content myself with knowing that I can never have absolute certainty about what happened on that day or on many others like it. In a comment that challenges our brain's correspondence with a video or tape-recorder, Loftus has said, "Memory actually works more like a Wikipedia page. You can go into your page and change things. But so can other people."

How can a system this unreliable have survived the evolutionary process? Precisely because what appears at first glance to be a *flaw* turns out to be a *strength*. Why would I remember getting struck in the head with a swing if the injury really had been inflicted on my sister? What possible benefit could be attached to such a defective understanding? Much as I hate to turn my back on all that I *think* I know to be true, I can make a case for my brain's possible errant behavior. I had been entrusted with my sister's well-being that day. How much better for my psyche and image of myself to believe that I had taken a blow while trying to rescue Gail rather than having to admit that I'd failed in my job as protector.

But wouldn't Gail also have had a reason to embrace a flawed narrative? Having acknowledged that possibility, the only way I can be comfortable with the unsettled nature of this memory is by arguing against her recall, thereby making a stronger case for my own. I begin by pointing out that Gail's brain had a very real motivation for deception. Remember that in my version of the story, I got hit by that heavy wooden seat when I was chasing after her. Had she not misbehaved in this fashion, neither one of us would have been exposed to danger. Dear empathetic Gail would have had a hard time accepting the fact that her determination to actually *play* at the playground had caused me far greater harm than any spanking ever had. Gail and I did not discover the disparities in our accounts of the swing set disaster until just a few years ago. By that time, there were no living witnesses to offer support for one version over the other. I told Gail that I had a scar on my scalp. She pointed to her cowlick and reminded me that it dated from our fateful day at the playground. Because, you see, that's the thing about false memories; once implanted, they are almost impossible to remove. We cling to what it is that we believe, and although Professor Loftus has convinced me intellectually that I could be mistaken, the emotional sense of my own truth remains unaltered. There are other lessons in this story. We don't know what we don't know. We remain unaware of the things we might have forgotten or, like the invisible gorilla, failed to see in the first place. Our brains are cleverer at their job than we may be willing to acknowledge. They tell us what we need to hear in order to hold to a coherent story about our past.

How do our brains compare to computers? As the above example makes clear, none of us would be satisfied with a computer that willy-nilly altered the data that were fed into it or chose to delete some files without bothering to inform us which ones it deemed irrelevant. I get furious when my email spellcheck—without waiting for directions from me—replaces words I'm attempting to type with others that completely change the meaning of the sentence. But at least I can *see* those errors. Actually, not even this last observation is completely true. Since I am inclined to read what I expect to be there rather than what actually appears before my eyes, I often don't catch those meaning-altering substitutions. Imagine a computer that went beyond simple word changes and replaced whole paragraphs or pages without alerting you that it had covertly deemed its take on the subject more likely to contribute to your success than your own analysis. I would never purchase a computer known for such shenanigans, but I seem to be stuck with a brain that can't be trusted to be fully on the up and up with me. So, would I willingly trade my brain for a computer? Of course not, even though computers are much better than I am at entering, locating, storing, and accurately retrieving data. Computers are far superior to human brains when it comes to computational skills, but there is a lot more to human intelligence than simple computation.

Part of the reason we have so much trouble recognizing *the value* in the brain's ability to do things such as implant false memories or ignore a gorilla on a basketball court is because we make the mistake of comparing our brains to the sophisticated mechanisms that we have designed to correct for and protect us from the brain's failure to provide us with completely accurate accounts of our lives, events, and the material world around us. This ability to *conceive of and design* amazing devices that perform a variety of tasks that we could not possibly accomplish without their assistance is one area in which our brains more than earn all the praise we bestow on them. How did we move from tool-making skills as basic as using a blade of grass to extract a termite from its mound to inventing the wheel and from that humble beginning to sprint ahead to launching exploratory missiles into space? A lot of it has to do with the evolution of that most wonderful of all human attributes: language. Without our skill with language, none of the other traits we hold dear would be possible. Language is the tool through which culture is spread. The spoken word enabled us to hand down from one generation to the next the knowledge necessary to survive in the wide range of environments that support human habitation. Once we had spoken language, it seems a natural progression to want to make the power of the word visible and permanent. The talent for symbolic art is associated with our earliest ancestors. It's not much of a conceptual leap to move from cave paintings and engraved tools to figuring out how to develop notations representing agreed-upon values in order to facilitate record-keeping and tabulations. Humans are the economic animal and were almost certainly involved in material exchanges early in our history. A mark struck in stone had the authority to spare our ancestors' early versions of "tape recorder fights" over who owed what to whom. Following the recognition that marks could substitute for thought and memory, the progression to the written word seems natural in populations above a certain size and with a need to communicate with those outside the circle of those with whom we had reliable, face-to-face connections. The written word is so powerful that once it came into being, denying enslaved and repressed others access to the instruction needed for mastery became a way of keeping them "in their place," a place that was designed to contain and prevent challenges to the authority of those at the top of the hierarchy. While I can't claim that humans are always an admirable species, no one who has witnessed the enthusiasm with which we adopt the latest technologies can be surprised at how difficult it was to restrict access to this new source of power, the written word, and how relatively quickly its widespread use made it possible to share knowledge and innovations of all sorts with those around the globe. I'm unable to look at the simplest of inventions without marveling at the human ingenuity and collaborative efforts that made them possible. I'm sufficiently impressed with the power of language to hasten our evolutionary journey that I will devote a subsequent chapter to the subject. But first, let's return to the difference between the human brain and the components of artificial intelligence that we've developed while using skills unique to *Homo sapiens*.

There are those who maintain that the human brain is nothing more than a very sophisticated computer and that our sense of self and consciousness is simply an illusion. Perhaps so, but the folks working in AI are having a difficult time designing a computer that can do all that our brains already do so effortlessly. For starters, designing programs that embody what we call common sense is proving to be a surprisingly difficult challenge. I suspect part of the problem is that those working in the field can't quite figure out how to unlink their one-to-one, computer/human-brain analogy from the target goal of developing a system of artificial intelligence that is consistently accurate. I don't happen to believe it can be done. The brain and a computer are different "animals" with differing goals. One focuses on survival and the creation of other humans; the other requires predictability and precision in the execution of tasks. As I hope I've convinced you by now, much of what the brain does so well is to ignore reliability in the service of survival and reproductive efficiency. Human brains do some of their most efficient work—at least if we accept that ensuring that we reproduce is the major directive through which evolution has shaped them—when they distort what we tell ourselves and each other about those to whom we are sexually attracted. I obviously part company with Darwin when he describes women as possessing an intelligence that is inferior to that characterizing men. Men and women are equally skilled at grasping and developing complex concepts and creative problem-solving. That we claim the superiority of one gender over another depending on the area of research—as in, for example, the STEM sciences versus language proficiency—has more to do with the subjective biases we bring to our assessments than it has to do with objectively measured skills. However, there is one very specialized field, courtship, where there are marked differences between how men and women behave and in their respective abilities to accurately perceive what is going on. Men and women may be equally smart, but basic biological differences between the sexes compel them to adopt differing strategies in pursuit of the universal evolutionary goal of reproductive success. It's time to consider the link between sperm, eggs, and brains.

Chapter 7: Sex on the Brain

As a species, humans are obsessed with sex. When a female chimp is in heat, all the males in the vicinity know it and become mad with desire. When she's not in heat, they kind of lose interest. Whether or not our unusual sexual innovations are to blame for the amount of attention humans pay to who's having sex with whom, no one can deny how preoccupied we are with keeping track of it all. Supermarket tabloids catch our eye with the promise of details on the latest couplings and un-couplings of celebrities we'll never meet. President Donald Trump's penis and payoffs to porn stars warrant discussion on national cable networks. It is generally assumed that men approach sexual encounters differently than the women who want to or are willing to have sex with them, and many of those differences have been empirically verified. The original division of labor began with the sort of specialization that is associated with sexual reproduction. It's a model that doesn't necessarily lead to equity in all the other realms of human behavior, and as those who struggle with these issues have learned, pretending it doesn't exist is not an effective way to guarantee justice. *The New York Times* recently (12/3/18) reported on a "study that revealed that roughly a quarter of people's views about gender equality are more complicated [than their publicly stated subscription to egalitarian beliefs], and differ regarding work and home. Most of them say that while women should have the same opportunities as men to work or participate in politics, they should do more homemaking and child-rearing." It would seem impossible to embrace both of these viewpoints without also being complicit in maintaining a system in which the pursuit of equality means that women are going to be working a lot harder than ever and even harder than men.

It's very difficult to override behaviors that have biological origins, but that doesn't mean it shouldn't be done. It simply means that we have to squarely face up to the fact that Mother Nature's idea of what constitutes a full and meaningful existence may be in conflict with our personal goals. Most men and women understand the relationship between the high cost of child-bearing and child-rearing for women and the need to address those costs in creative ways both within their personal relationships and through social policies that enable women to achieve their full potential in their engagement with the wider world. The continued existence of a wage gap between men and women is but one indication that the problem has not yet been fully solved.

This wage-gap differential is more easily measured than other indications of gender inequality, and it shrinks wherever there is reliable and affordable child care. All workers benefit from policies supporting a realistic number of days off for personal and family emergencies, employment opportunities that are open to scheduling flexible hours, and accommodation of job-sharing arrangements. These things benefit both men and women, and both should be encouraged to use them in ways that enhance productivity at home and at work. Unfortunately, there is too often an unarticulated but very real bias against men who avail themselves of such programs. The sad truth of the matter is that we tend to continue the inherent *imbalance* in the original division of labor imposed on us by a system of sexual reproduction long after it is physically necessary. The higher cost to women in the creation of a new human being is most obvious and most difficult to alter in the nine months between conception and birth. However, once a baby is born, the costs and benefits of child-rearing can be more equitably shared. We are a social species. Our deepest and most emotionally significant bonds are with our family members. Everybody benefits and is happier when nobody feels exploited or underappreciated. This is as true at home as it is anywhere else.

Early sociobiologists and current evolutionary psychologists give the nod to the cheap sperm/expensive egg sound bite as a way to quickly explain why females are the more selective of the two sexes, but the scarcity of eggs relative to sperm is just the beginning of the story. Throughout her life, a woman's physical investment in the creation of another human being is much higher than a man's. Once she becomes pregnant by one man, her baby-making equipment, busy with other matters, is effectively frozen against additional impregnation by this man or any other for nine months. No further pregnancy can occur until she has finished work on the one in progress. And if she breastfeeds, ovulation and its associated reproductive availability can be delayed until she is no longer nursing an infant. The man responsible for the pregnancy has no such physical constraints. He can continue to impregnate other women throughout those nine months. Social and emotional prohibitions may discourage him from availing himself of this physiological opportunity, but it's difficult to argue that the cost of child-bearing and rearing is equal for men and women. Since the female has already made a heavy investment in the months preceding birth, she is highly motivated to protect that initial expenditure by continuing to invest heavily, putting her baby's needs ahead of her own, and not jeopardizing her child's welfare in a postpartum refusal to contribute more than fifty percent of the physical and emotional investment now required to keep this little one alive. She's already contributed far too much to risk the potential losses that might be incurred by going on strike in pursuit of an equitable parenting arrangement. It is, of course, about so much more than a fifty-fifty distribution of tasks. It doesn't take long for both mother and father to fall completely in love with this miraculous little human being. One father told me, "My daughter is only two weeks old and I already know that I love her more than I've ever loved anyone in my entire life. I wasn't prepared for this. I would do *anything* to protect her and keep her safe."

Just as the father quoted above was unprepared for the extent of his devotion to his daughter, so are both mothers and fathers pretty much unprepared for the all-consuming amount of work, energy, love, and commitment involved in rearing a child from conception to adulthood. If there were a way to possibly know everything that is involved, would the vast majority of us still go ahead and have children? It's an enterprise with both high costs and high rewards. Our social desires are sometimes in conflict with a biological imperative that puts a heavy emphasis on reproduction. Even people who are intellectually certain they do not want children to exhibit behaviors that make them vulnerable to assessing romantic partners for signs of commitment and fidelity that aren't necessarily functional if an unencumbered lifestyle is the actual goal. Without reproduction, there would be no evolution. This is what we have been shaped to do. This is one of the primary things our brain is good at making happen. The high physical, emotional, and material demands associated with having children are the cost part of the cost/benefit equation. How is it that so many of us decide to become parents? In particular, given the disparity of investment, how is it that so many of us decide to become *mothers*? We do so because there are also benefits, rewards that trigger the pleasure centers of our brains in ways that blind us to the potential consequences realized *tomorrow* of actions we undertake *today*. The brain may not be a completely rational instrument, but it does know how to make the behaviors leading to reproduction a lot of fun. What is more delicious than falling head over heels into someone's arms? What feels better than sex? Nursing a baby floods the brain with oxytocin, a hormone associated with bonding, trust, and love, and one that is also released during orgasm. Can anything in the world compare to the love we feel for our children? Our species would not continue to exist if we failed to reproduce. However, given the physical differences between men and women and the disparity between the investments required of each in order to ensure that they succeed in getting their own genes into the gene pool of the next generation, men and women have had to evolve different strategies when playing the mating game. The cheap sperm/expensive egg explanation may be nothing more than a starting place that neglects a full accounting of all the differential costs that women experience as compared to men, but it serves nicely as a metaphor. The goal for both men and women is the same: get as many of your own genes into the gene pool as possible. If we ignore the full range of factors associated with parenting and focus solely on sheer numbers, it's easy to see the origin of stereotypes about differences in male and female approaches to sex. For men, the strategy is pretty clear cut: get as many women pregnant as possible. Don't worry too much about whether the children of these pregnancies live or die. Impregnate enough women and the odds are in your favor. For women, it's a whole different game. Maximizing quantity over quality doesn't work for us. We have to pay more attention to whether or not any child resulting from a sexual encounter with this particular man has a good chance of surviving into adulthood.

So, one strategy that women employ is to "mate *up*," to fall for the guy with the best available genes and enough potential as a wage-earner to inspire visions of life-sustaining investment in his offspring. Our expensive eggs caution us not to waste them on anything but high-quality sperm. We want our children to be born healthy. In addition, human infants are relatively helpless and require years of intensive parenting before they are able to make it on their own. Their chances of survival are greatly increased when they have the love and nurture of two parents. So women not only look for men who are near the top of whatever hierarchy defines success in their particular culture, but they also seek men who appear to want a long-term relationship and have the economic means to make continuing investments in a child's well-being. Sometimes the offer of financial support is enough, but more often, women need to believe in the promise of happily-ever-after. The stereotypes associated with these differences in mating strategies can give both men and women a bad rap. Most men are not careless philanderers. Most women aren't gold-diggers. The focus on measuring a man by his material assets is born of the biological dictate to do all we can to ensure that any child resulting from this encounter will have a reasonable chance of having access to the food, shelter, and education it needs in order to reach successful adulthood.

In the previous chapter on the brain, I noted the ways in which an ability to deceive ourselves as to our intent—past, present, and future—could enable us to survive and get on with our lives when more accurate assessments might suggest the futility of it all, but I glossed over the ways in which our very skill with language can work to produce a reproductive outcome that may be at odds with the expressed desires of the individuals in question. The fact is that our way with words can be employed in the service of both self-deception and deceit that on the surface doesn't look to be particularly adaptive. Our capacity for language and the special nature of human intelligence evolved together. It's impossible to truly tease the two capabilities apart. In combination, they've produced a species that has transformed the face of the earth. What *Homo sapiens* has achieved in terms of technology and artistic expression is so different in degree from what other species have done that it is almost impossible to comprehend. But all the scientific progress made possible by advanced theories and mathematical precision, the skyscrapers, space explorations, symphonies, and astonishing works of literature and visual art should not blind us to the fact that we are still animals, animals that evolved under the same directive to reproduce that has shaped every living thing on the planet. We may be good at reconfiguring the world to serve our immediate needs and desires, but the one area where Mother Nature cannot allow us to seize control is in the mandate that we make copies of ourselves. If we become so clever that we figure out how to outsmart this evolutionary push, we will have effectively guaranteed our own extinction. Reproduce or die. It's that simple.

I argue that sexual selection drove human evolution when women chose men with the skills that made it possible for us to survive in a world to which we had not been adapted by millions of years of natural selection. The end-to-end chromosomal fusion that served as our speciation event came with both advantages and disadvantages. Any mutation of this magnitude was bound to have been accompanied by some deleterious side effects. However, the plus side of that dramatic shift in genetic architecture resulted in a neural restructuring that set the stage for language acquisition. We weren't created with a full-blown capacity for speech. We didn't spring from that chromosomal fusion with words in our mouths and a language organ in our brains. However, our ancestral Eve *was* born with ovaries that knew how to keep a secret. The mutation that gave us concealed ovulation and continuous sexual receptivity was part and parcel of the reduction in chromosome pairs that set us forever apart—both physiologically and behaviorally—from the great apes. Eve actively drove the rapid evolution of the creature we have fondly dubbed *Homo sapiens* when she employed the option now at her disposal to select for any tiny skill that moved us toward verbal ability. Our way with words and our clever brains evolved together. It all happened so quickly that human evolution can't be explained by falling back on the slow adaptations associated with natural selection. Sexual selection is the only force logically capable of producing the combined traits responsible for human exceptionalism.

But no matter how smart we became, any woman who figured out how to override a push toward reproduction effectively removed herself and any genes capable of making such a decision from the assortment that got carried into the gene pool of the next generation. Language ability can be a double-edged sword. It enables us to share the information that makes cultural adaptations possible, but it can also be used to create the little white lies we use to fool ourselves and others. We women are especially vulnerable to telling ourselves the stories we need to believe in order to, wittingly or unwittingly, expose ourselves to the possibility of pregnancy. For us, sex on the brain is tightly linked with the brain's language ability. For men, sex on the brain seems to be a bit less complicated by internal debate and a bit less reliant on self-deception. A man doesn't have to kid himself about his motives. He wants to have sex. Where he can get a bit careless is in a failure to make certain the woman involved is not kidding herself about his emotional commitment—or lack thereof. This need to sort fact from fiction provided a selective pressure for ever greater brainpower, creating a positive feedback loop that made us who we are in a rather impressive evolutionary sprint. In one of the early chapters in my 1980 dissertation, I made the following observations when considering the ways in which language contributed to the rapidity of human evolution by amplifying the power inherent in sexual selection.

> One of the big advantages of a system of language is that it makes the future conceptually available to us. Without words for tomorrow or next month, or next year those time frames remain beyond our grasp. A female who is in the process of deciding on a sexual partner makes her decision in the here and now, but the impact of her present actions will be experienced for years into the future. [Earlier in these pages], we saw this concern with the future take the form of desire for commitment. "He took my name and address before we went to bed, and I think that kind of helped." One woman was persuaded to have sex with a man who had only to tell her, "We're going to be friends for a long time." A male who is able to speak of the future and pledge that he will remain with the woman into that future has a distinct advantage over a male who can make no such claim. Language further aids as a screening device. A woman does not have to witness a man engaged in all his various activities if he is able to convey them to her through conversation. Of course, the development of language also introduces the possibility of sham commitment. The ability to lie was born with the ability to speak. This leads to further selection pressure for intelligence. The push for the development of intelligence in males was that females were looking for abilities that would aid a male in his quest to accrue resources. The push for the development of intelligence in females was, in part, the need for cleverness in the detection of sham behavior by males. Language made all this possible. It also enhanced the development of a social animal. [Human] survival is very dependent on the nature of [our] social networks. Certainly, the bonding process is strengthened by a whole range of physiological reactions. Pheromonal processing may play a larger part in human bonds than anyone has yet supposed. The continuous sexual receptivity of human females has been credited with being a factor in the maintenance of systems of marriage. That said, [one woman] speculated that it was probably not sex but conversation

> that kept most marriages together. Her theory was that sex was a very poor way for a woman to permanently ensnare a man, and she pointed out how common it was for a woman to have sex with a man who then abandoned her. But the men with whom women have intimate conversations? Those men remain trusted friends for years. This woman looked at her marriage and decided that because her husband could talk to her about his innermost feelings in a way that he could not with either men or other women, their marriage bond was strengthened. The kind of trust it takes to accomplish this is established over years and makes a man less vulnerable to compulsive infidelity than if the marriage bond were based on sex alone. Why do men need women? Perhaps they need them as much for conversation as they do for sex.

What are some of the ways in which the human facility for language might enhance the reproductive success of those who demonstrate skill in this area? I've pointed out that a man's promises of a committed, until-death-do-us-part bond gives him an edge over someone who lacks this grand vision, as can his ability to describe his own past achievements, if delivered convincingly and without off-putting braggadocio, serve to inform a woman of his status and desirability. Other than enabling a woman to take the measure of a man and get to work weaving the verbal ties that bind, how does language contribute to *her* reproductive success? Sometimes simply by making sure she gets pregnant in the first place, and in order to do that, the brain has been known to resort to a bit of trickery, trickery made possible by employing the words we use to deceive ourselves.

Here's the catch. Although many women are drawn to the man at the top of the hierarchy, not all of them are not going to have a deeply committed relationship with him. Speaking from the female point of view, the evolutionary pressure on women is to get pregnant by a man with healthy sperm and a demonstrated measure of social and material success. In a mostly pair-bonded species such as ours, one which is characterized by a significant amount of *paternal* investment, women are going to fall in love with smart, healthy, accomplished, generous men who demonstrate a willingness to form lasting bonds with them, and although these traits are appealing *because* they have in our evolutionary past contributed to the survival of the children fathered by such men, the traits themselves trigger desire in women regardless of whether or not they actively want to get pregnant. Since there are a lot of women drawn to alpha males and only a relatively few men at the top, not all women are going to successfully achieve the lofty goal of snaring one. So here's the dilemma Mother Nature faces: she wants women to be choosy but not so choosy that they fail to get pregnant at all. In order to solve the inherent contradiction in the message to women, M.N. turns to the trickery made possible by our lying brains. The directive to women appears to be: "When you can, find the perfect man, and when you can't, invent him." This reproductive charge is one of the reasons the human brain has evolved to be so good at self-deception. It guarantees a steady supply of new human beings. Women are really good at inventing men. Women are so skilled at lying to themselves, at self-deception, that it wasn't necessary for men to have achieved a matching talent for *actual* romantic deception. I'm old enough to have spent my reproductive years in an era when pen-to-paper letter writing was a primary means of communication. Examining the letters sent to us by former lovers is not nearly as reliable a means of exposing this talent for deceiving ourselves as is asking trusted female friends to dig out the letters we sent them when we were in the first blush of love. It can be a bit embarrassing to compare the traits we ascribed to men newly on our romantic horizons to the assessments we made of those same men a few years later. That is not to say that we made bad choices. Very often, the traits that first inspire us to fall in love are far less reflective of the man's true worth than are the less heart-stopping but more endearing, and *enduring*, ones that we come to recognize and value in the fullness of time. However, there is a warning in this for all women. Beware of your ability to ascribe traits to men that they may not actually possess. We can save ourselves a lot of heartbreak by acknowledging in advance that we have tendencies in this direction. One of the ways in which women identify the so-called alpha male is by noting how appealing he is to other women. Be on high alert for the trap in all of this. If scores of women desire him, the odds of your being the one who actually captures his heart are not necessarily in your favor. One of the things that triggers a woman's sexual desire for a man is the sense that *he* desires her. After all, a man who doesn't communicate genuine desire but appears to simply be going through the pro forma behaviors that he thinks will lead to sex is not a very good candidate for happily-ever-afters. The longing in a potential partner's eyes can help a woman believe that he will remain true. The appeal is especially intense when the man in question is a prominent figure. I found something heartbreakingly vulnerable in Monica Lewinsky's description, in a clip promoting a recent A&E documentary series, of her helplessness in the face of President Clinton's magnetism, "It's not as if it didn't register with me that he was the President," she said. "Obviously it did. But I think in one way, the moment we were actually in the back office for the first time, the truth is that I think it meant more to me that someone who other people desired, desired me. However wrong it was, however misguided, for who I was in that very moment at 22 years old, that was how it felt." Monica Lewinsky is not alone. I've talked with enough women to believe that most of us, whether we're willing to admit it or not, would have responded in exactly the same way if we'd found ourselves in a similar situation with the handsome, charming President of the United States. Notice the passive coyness in my phrase, "found ourselves." It's another example of our skill with self-deception. We are very able to forget, overlook, and ignore all the behind-the-scene machinations we actively pursue in order to *suddenly* find ourselves alone with the man with whom we've "suddenly" fallen in love.

Chapter 8: First Words: The Evolution of Language

Imagine a scenario where instead of finding ourselves in the Garden of Eden, we—a brand new species—wake up surrounded by a hostile environment, one in which we are poorly equipped to survive. If the dramatic speciation event which gave us a shove down the path leading to a creature with all the characteristics we associate with *Homo sapiens* began with an end-to-end chromosome fusion resulting in what is now human chromosome 2, the moment of our actual birth would have been less traumatic than our subsequent adventures might prove to be. After all, mothers in most species are very good at loving and taking care of new little copies of themselves, even when those copies are not particularly exact. Picture a chimp-like primate contemplating the hairless mite she is attempting to get to suckle. Once prototype #1 latches on, a flood of oxytocin washes over its mother's brain and transforms her perception of the infant at her breast into the most beautiful little creature she has ever seen. The newborn nurses vigorously and grows, but its rate of development is unusual and seems to lag behind that of its peers. Even though this child of hers doesn't scamper off on its own quite as quickly as other youngsters in the troop, by the time those athletic deficiencies become apparent, Mama is already so deeply in love that her continued investment in little Adam or Eve simply intensifies. I don't think verbal skills would have been manifest in early-edition humans. Those no doubt took some fine-tuning at the hands of both natural and sexual selection, but I'm willing to speculate that hairlessness and ineptitude with knuckle-walking were not the only system-wide changes that were immediately expressed. The opposable thumb is a possible candidate and that, combined with slight changes in intelligence, would eventually give patient zero an advantage over its differently endowed companions.

Fast forward to adolescence. Are there appropriate reproductive partners anywhere in the vicinity? If mutations of the sort that lead to end-to-end fusions can be triggered by an abrupt environmental event, there might even be another primate that closely resembles our teenaged misfit. In that case, assuming we've got both a male and a female model, the gene recognition theory of attraction would kick in and give the whole speciation process a jumpstart. But suppose this newly naked ape is one-of-a-kind? Then we have to fall back on the allure of the dramatically different stranger. But this time, the stranger has more to recommend him than simple novelty. He's really good at picking up thick-shelled nuts that fall from a tree he has difficulty climbing. Since his old-model primate companions ignore this bounty in favor of fruits they harvest from treetops and are unable to crack those nuts anyway, he uses that brain of his to figure out that the easiest way to access the flesh is to find the perfect rock and smash it against a firm surface. The ladies love this demonstration of his prowess, and when he picks up a piece of the exposed meat and slips it into the mouth of one of the females who have gathered 'round to watch, courtship feeding has established itself as an effective aphrodisiac.

Okay, I confess that recall of my interviews with women has influenced my speculative account of the origin of our species. I find it impossible to put out of my mind what one woman told me regarding her decision to have sex for the first time with the man she later married, "He gave me a bite of his peanut butter and jelly sandwich and the next thing I knew, we were in bed."

I encourage readers to explore their own hypotheses regarding the mutations and adaptations that they think could explain our establishment as a distinct species. Mine begins with an episode of end-to-end chromosomal fusion. Manuel Ruiz Rejón, the Spanish researcher whose work with the fusion that could be either our original sin or our path to glory—depending on your assessment of the ultimate outcome—notes that the genes in the region of the fusion that are expressed more intensely in our species than in non-human primates are associated with the brain and the gonads. Therefore, any story I tell about how it all began focuses on sex and the brain. I've already given the nod to brainpower, but please don't forget the importance I ascribe to the unique-to-humans combination of continuous sexual receptivity and concealed ovulation.

Before moving forward with theoretical accounts of how we ended up becoming *Homo sapiens* rather than one of the now-documented evolutionary dead-ends in the hominin line, consider the possibility that at least some of those other, less successful, variations on an earlier model also may have been the result of chromosomal fusions. Lucy, officially designated *Australopithecus afarensis,* was not an ancestral human. She represented a line with an upright posture and a chimpanzee-sized brain that went extinct about 140,000 years ago. Researchers have found 250,000-year-old fossils from a South African species, *Homo naledi,* that also had a chimp-sized brain. And then, of course, there is the tiny *Homo floresiensis,* hominins who lived until 60,000 years ago on the Indonesian island of Flores. We don't know how many chromosomal pairs these early hominins had, in part because DNA degrades over time. In 2013, researchers sequenced the earliest sample of hominin mitochondrial DNA, that of a 400,000-year-old femur found in a cave in Northern Spain. The analysis was made possible only because unique climatological conditions in the cave where the femur was found preserved the genetic material. Advances in genetic analysis are proceeding so rapidly that I'm confident it will someday be possible to answer some of the questions I raise. If evidence of chromosomal fusion is found in these extinct branches of the hominin line, I suspect they will involve parent chromosomes other than the ones whose fused reduction subsequently led to modern humans. My guess would be that concealed ovulation was not part of any genetic package resulting from fusions other than the one that led to our chromosome 2. Our line of descent diverged from the one containing chimpanzees and gorillas, meaning that we are not the direct descendants of either of these species. Our most recent shared ancestor with chimpanzees was between 5 and 7 million years ago. The fusion resulting in chromosome 2 is estimated to have occurred between 0.75 and 7 million years ago. We modern humans, also known as *Homo sapiens,* might be no older than 300,000 to 330,000 years. Only 600,000 years ago, we shared an ancestor with Neanderthals and Denisovans. In evolutionary terms, we are a relatively young species. If we make sexual selection part of the game plan, the twig that developed from that initial chromosomal fusion has had more than enough years of Darwinian selection pressures in which to work its magic and produce a creature capable not only of pondering its *past,* but also of conceptualizing and designing spacecraft that could, in the *future,* enable us to colonize other planets.

Before moving on from the humble beginnings made possible by chromosomal fusions, I want to introduce one further speculation for consideration. Early Modern Humans co-existed and mated with both Denisovans and Neanderthals. In an earlier chapter, I pointed out that not all chromosomes are created equal. Some are more vulnerable to nondisjunction events than others. Perhaps the same is true of chromosomal fusions. In that case, an environmental occurrence such as a major solar flare could have triggered a vulnerability to mutational consolidations resulting in the sort of reduction that almost certainly played a central role in our own genesis. I think it's possible that species that were already closely related could have experienced separate but equal and near-simultaneous fusions. In other words, Early Modern Humans, Neanderthals, and Denisovans could have independently, and in different regions of the world, all experienced the consolidation of great ape chromosomes 2A and 2B into the single chromosome 2 that later made breeding between these three groups possible.

Of all the amazing things humans do, nothing is more miraculous than our linguistic ability. Without this way with words, we would not be able to consider either incidents from the past or design and plan for the future. It's not just our personal history and experiences that we remember. The encoded discoveries of others who died long before anyone now living was born provide the foundation that guides current research. Our technological prowess is totally dependent on symbolic representations that enable us to draw on established knowledge and share our conceptual frameworks and current discoveries with others. We are a social species that can accomplish far more by pooling our expertise than we could ever manage on our own. We specialize in divisions of labor and mastery. While typing this, I look around my study. Other than the arrangement of the words that appear as if by magic on the screen in front of me, there's almost nothing in the room that I have made or could have made all by myself. The words I employ are themselves spelled out in the language of my parents and are something I had to learn from others. Not even the ideas are wholly mine but have been shaped and triggered by decades of intellectual engagement with those either living or long dead and who inhabit/inhabited geographic areas both far and near. It's impossible to calculate the combined total years of human effort and ingenuity that produced the insights and abundance of material things that fill the small room in which I write. In addition to shelves and leaning piles of books, I am surrounded by objects so mundane that I hardly give them a thought. A mug that holds a pair of scissors, a brush to clean the keyboard, and an assortment of pens and pencils. A lamp. File cabinets. A desk. Framed photographs of loved ones. Binoculars on the windowsill. The glass in the skylights. Closets filled with items too numerous to count. The condo conversion in which I live was once a school. Who made the bricks in the walls? Who quarried the ancient Monson slate that shingles the roof above me? Who shaped the beams? Who felled the trees? What architect calculated the angles, the pitch from ridgeline to rain gutter, the precision with which they fit together and have held together for all these hundred-plus years? Who made the thermostat that guarantees warm air circulating around my feet on a cold, winter-in-New-England day? I've barely scratched the surface of the endless stuff in my immediate space, and I'm already exhausted by the attempt to come up with a list. Even if I'd been born with the inventive genius of an Edison or Einstein, it would not have been possible for me to create the wealth that I dare to call mine. Not in one lifetime. Not in a hundred lifetimes. Not in a thousand lifetimes. That's what language makes possible. That's what mechanical prowess makes possible. That's what the swapping of specialized skills makes possible. That's what cooperative human endeavor makes possible. We are the social animal. We are the technological animal. We are the economic animal. Above all else, we are the language specialists of the animal kingdom.

Language is what makes our unique brand of intelligence possible, and any discussion of it must begin by making a clear distinction between language and communication. Many species, even plants, have complex systems of communication. I suspect that as we learn more, we will come to realize that every living thing has some system of communication. However, humans are the only species with language. It is language that makes us unique, but not necessarily superior. I balk at claims of human exceptionalism that are based on scores derived from tests loaded with questions that skew the odds in our favor.

Until I read Derek Bickerton's book, *Adam's Tongue*, I really hadn't tried to come up with a coherent definition of language, despite the fact that I'd long sensed the importance of the future tense as a verbal behavior that might be subject to selective pressure. In the research I conducted, I was considering traits that would be especially responsive to sexual, as opposed to natural selection. As I've mentioned before, a man who can promise future devotion has a leg up on one whose verbal skills restrict him to the present. Bickerton's insistence on displacement—*the ability of symbols to refer to things outside the immediate here and now*—as a feature necessary to any definition of language nails it for me. Unless displacement is a component of other species' communication skills, we can't describe what they are doing as using language.

On what mutations did the processes of natural and sexual selection operate in order to favor the development of language skills? This is a question at the heart of understanding how language evolved. Bickerton early on makes the point that we should try to figure out what human traits are different from our nearest primate relatives in order to get an idea of the uniquely human experience/niche/condition that would have favored the selection of any novelty facilitating the development of language. He decides that scavenging for large dead animals on the savanna is the first step in the development of language. I disagree. First of all, species such as hyenas also engage in this kind of behavior, and while they may have good systems of communication, I see no evidence of a move toward language. In addition, humans can highjack the communication skills of turkey vultures, for example, in order to locate a large dead carcass. We don't need to talk with each other about where the dead wooly mammoth is. We can either point to the buzzards or—if they are out of sight—gesture, mimicking their circling flight pattern and pointing in the right direction. These kinds of informational exchanges happen in the moment and don't require a Bickertonian displacement in order to be an effective means of communication. Bickerton then goes on to argue that we should not limit our search for language-enabling traits to primates and describes the signaling used by some in the insect world to direct other members of a beehive or ant colony to sources of food that are out of sight. He feels such behavior offers evidence of the displacement property necessary to language. However, since ants and bees are haplodiploid species, this explanation doesn't work for me. All members of a colony are so closely related as to function almost as a single organism. Therefore, I can't accept their messaging as evidence of displacement ability but think it is as more akin to the kinds of hormonal, electrical, and chemical signaling that occur within a single individual of any species.

For me, the most fruitful starting point is to look at the thing Bickerton mentioned initially, but then didn't return to in any meaningful way: how humans differ from our closest primate relatives. There are a number of factors unique to the human condition that could be employed in the selection for language and our special system of intelligence. Many of the standard (concealed ovulation is a *non-standard* candidate) anthropological explanations come into play here: upright posture (which frees the hands for gesturing), opposable thumbs (allowing for efficient tool use and invention), the fact that we're a social species (gossip helps keep track of who is doing what, even when out of sight), tool use (enabling niche creation and complex engineering of the environment), hunting, scavenging, and foraging skills (much enhanced by efficient communication between members and driven to a whole new level once there is capacity for displacement), ethnobotany (knowing what is good to eat, which plants will kill you, and which will cure you enables the exploitation of novel environments), artistic expression and musical ability (both dependent on skill with symbols), extended infant dependency (the human brain continues to grow and develop postpartum; plasticity favors learned adaptations to unique environmental conditions), continuous sexual receptivity (both a pair-bonding device and a means of persuasion employed to solicit investment in a woman's children), and the physiological structure of the mouth, throat, vocal cords, etc. (a necessary but insufficient component of spoken language). Selection based on any of the above traits can be argued to contribute its piece to the development of a complex system of language. But while no single trait may be sufficient on its own, there is one that is *necessary*, one without which none of the others would have had an opportunity to work their magic. For this one, we may need to look no further than the slight craniocerebral changes that accompanied the initial fusion, which I identify as our moment of birth as a species. What two areas of the genome were affected and intensified by that fusion? The brain and the gonads.

Interestingly enough, neither of the mutations that I put forth as central to human evolution would have resulted in visible changes. Concealed ovulation, by its very definition, takes what was flamboyantly advertised and makes it covert. Even continuous sexual receptivity waits until adolescence to manifest and is most effective when practiced with a sleight of hand that masks its intent. The change that I nominate as being absolutely *necessary* to language development occurs in the brain and would not, I think, show up even on an

MRI. Following that initial chromosomal fusion, the structure of the brain would not have changed, but something would have happened to allow the brain to accommodate a novel kind of plasticity. The cortical areas of the brain would remain architecturally the same, but the wiring, the ease with which neurons fired, transcribed, and communicated information, would have undergone a transformation that might best be described as allowing a *co-opting* of old functions in order to put certain regions to new use. A change of this sort would enable the system of encoding necessary to the production of symbolic expression that is unique to humans. Some old functions could have been so vulnerable to takeover that ancient proficiencies were excised and, if not lost altogether, at least buried so deeply as to be difficult to access without practiced use. One that comes immediately to my mind is navigational skill based on an almost-innate mathematical sense of geometry and physics. Others, such as regions used in the processing of aromas and chemical and pheromonal messages, could—as a result of the system-wide impact of the fusion—been shifted to other areas of the body. "More than 15 of the olfactory receptors that exist in the nose are also found in human skin cells." When and how did that happen? And given the evidence that it did, is it any wonder that touch is such an important component of the courtship behavior of this newly hairless ape?

No one has written a more compelling analysis of the evolution of language than Terrance Deacon, neuroanthropologist and member of the Cognitive Science faculty at the University of California, Berkeley. I read his book, *The Symbolic Species: The Co-evolution of Language and the Brain,* four or five years before I became aware of research suggesting the importance of chromosomal fusion in the search for human origins. I recently made my way back through all the notes I'd scribbled into the end pages of my copy and was stunned at how pertinent many of the insights he'd had while writing a book that was first published in 1997 were to current evolutionary debates. I don't have a background in neurology and best retain new information by finding a place for it within existing theory in my own field. There is so much about the structure and function of the brain that I didn't and still don't know that I found his book to be one of the most challenging and exciting books I had ever read. I completely trust Deacon's authority and hope that my layperson's takeaway from his work is not overly simplistic. In my reading of the book, I naturally focus on observations that speak to my own special interests. In doing so, I will almost inevitably distort his treatment of the subject. It's impossible to distill Deacon's complex understanding of a complex subject into a few sentences. I urge readers to go directly to the source. You'll come away filled with awe at the almost impossible-to-imagine evolutionary intricacies involved in the development of language abilities and of all that this human skill makes possible. *Homo sapiens* starts here.

Deacon rejects what are called hopeful monster theories, "the evolutionary theorist's counterpart to divine intervention, in which a freak mutation just happens to produce a radically different and serendipitously better-equipped organism." By speculating that system-wide changes would have accompanied the chromosomal fusion that biologically defines us as a separate species from our nearest primate relatives, I am indulging in a hopeful monster scenario of my own. However, since I happen to believe that not all the restructuring associated with such a mutation would have been beneficial, mine falls into the *not-so-hopeful-monster* category. Some of the modifications almost certainly would have imposed a handicap on Prototype #1. In order to survive in an environment to which this new species had not been gradually adapted, early humans would have had to exploit at least some of their new features in novel ways. As is often the case, the "workarounds" needed in order to function in the face of a disability can lead to the development of outstanding strengths in other areas. I speculate that, at the moment of fusion, some physical or chemical change either in the sheaths surrounding the neurons or in the neurons themselves may have allowed a different sort of information to be processed and recorded, information related to symbolic meaning. Terrence Deacon, please forgive me for what I am about to do. You may not be willing to accept my starting place here—a dramatic, almost overnight change in some aspect of brain chemistry—but it is grounded in one of my takeaways from your book, a belief that language ability co-opts and is encoded into existing brain structure, modifying it in the process.

From the first moment I became interested in the relative speed of human evolution, I sensed that understanding the evolution of language would help unlock the riddle of our sprint to what some would consider the top of the pyramid of life and what others might view as our race to destroy the planet. In the notes I took at the time I first encountered Deacon's book, I scribbled my understanding of this observation in the flyleaf. "Brains evolve slowly. Language does not. Cultural evolution proceeds much more rapidly than biological evolution." Of course, although I now believe that some change in the brain *was* dramatic and immediate, Deacon was fairly clear in his position. "Biological change is vastly more slow and inflexible than language change. Brain evolution takes place on a geological time scale. Even slight changes probably take hundreds of thousands of years to become widely represented in a species, and the basic architecture of the brain has been remarkably conserved since the origins of vertebrates. Languages, on the other hand, can become unrecognizably different within a few thousand years." I agree that the basic architecture is not dramatically different, but it was my early reading of Deacon's description of the ways in which language ability takes advantage of existing brain structure that had primed me, on learning of that chromosomal fusion, to envision us as being born of a spectacular event that was accompanied by abrupt changes that made it possible for early humans to suddenly use an old brain in a brand new way.

My Not-So-Hopeful Monster had a lot of handicaps to overcome. In a stroke of luck that falls into the winning-a-record-breaking-lottery category, one of the other transformations that accompanied this system-wide shift in genetic expression was the elimination of flamboyant estrus in favor of concealed ovulation and continuous sexual receptivity. Prototype #1 may have been hairless, doomed to upright posture's lower-back problems, and lousy at climbing trees, but she had two new features that could be employed in overcoming those handicaps: a brain with some unusual wiring potentialities and reproductive innovations that ramped up the power inherent in sexual selection and enabled it to go from good to unbelievably awesome. All of this set the stage for the relatively rapid joint evolution of language, and *Homo sapiens*, a creature whose growing symbolic capacity helped it figure out how to put a new twist on the way the adaptation game is played. Instead of having to wait for selection pressures to shape us to the environment in which we found ourselves, we were beginning to figure out how to use our language-enabled technological skills to shape the environment to our own needs.

One reason our brains are so good at convincing us of "truths" that enable survival and reproduction but that don't necessarily meet camera/tape-recorder/computer standards of external objective reality is because at least some of what language enables is deception. However, not all deception is language-dependent. The combination of concealed ovulation and continuous sexual receptivity can be interpreted as starting things off with a physical falsehood that operates independently of words. While not themselves language, these linked reproductive innovations can be employed to hasten the development of the kind of verbal finesse that is sometimes used to hide the truth. After all, what is concealed ovulation but a kind of secret-keeping that comes close to prevarication? When old-model male primates had sex with a female, they "knew" her sexual receptivity had a pretty tight correlation with her reproductive receptivity. In other words, if you managed to fight your way through the crowd and have sex with her, there was an unfertilized egg just waiting for some lucky sperm to change its status from egg to embryo. New-model primates come with no such guarantee. In fact, humans are much better at not having babies than at having them. By breaking the link between sexual readiness and reproductive readiness, females gained a lot of power to entice males into forming pair bonds and making the kinds of *paternal* investments that increase any baby's chances of survival.

To be human is to have a fully developed capacity for disingenuousness. Our original sin was an ability to lie that was born of a sexual innovation that didn't need the seductive charms of a serpent to coax it into existence. Linguistic virtuosity was driven by an arms race between the sexes that began in biology and ended in literature, technological skill, art, music, resource accrual far in excess of anything needed for mere survival, and—yes—religion, philosophy, and the scientific method. It is language that makes the coiled intricacies of our deceit and self-deception possible. Could it be that mathematics is the only complex symbolic communication that is unblemished by falsehood? When we seek to establish replicable and agreed-upon truth, we rely on the scientific method to keep us honest. The levels of accuracy we acquire through controlled observation and experimentation are the foundation of technological innovation and widespread advances in human understanding. We grasp the importance of establishing objective truth but tend to conflate intelligence with devious intent and are willing to grant the former to other members of the animal kingdom when we observe them practicing the latter. Some of the proof we turn to when documenting mental capacity in other species is exhibited through deceptive displays, such as when one raven pretends to cache food in order to prevent his fellows from figuring out where the surplus is really hidden. This sort of behavior enables researchers to claim that crows and ravens have a theory of mind. While there is no denying that corvids are highly intelligent, I wish we granted equal time to their tool use and skill with facial recognition, and were a little less impressed by their knack for deception, a trait at which humans are the undisputed champions.

Much of the increase in brain size associated with human evolution is lodged in the prefrontal cerebral cortex. Our "new" brain did not replace the old brain but is built upon that original model. The neocortex, which probably contains much of our language-enabling network, is an enhanced edition established on top of more primitive regions of the brain. Humans retain what some researchers refer to as the reptilian brain stem. I believe that many animal species communicate with one another through chemical signals called pheromones that are lodged in and are dependent on these older structures. This system of communication is inherently honest. Language is not. Language not only allows for displacement but also enables multi-layered cognition and deception. At any given moment, a person can be saying one thing and thinking something entirely different. Just imagine the social chaos that would erupt if all our innermost thoughts were suddenly apparent to everyone in our social group. Language enables deception. Pheromones don't lie.

I like to believe that humans still have a working reptilian brain that enables us to tap into the inherent truthfulness of pheromonal communication. It's a backup system that operates at a subliminal level to help us access the trustworthiness of others. Recent research identifying olfactory receptors located throughout the human body leads me to speculate that these receptors are used in subliminal pheromonal communication. Are the things that other humans *say* they are feeling accurate approximations of what they are *actually* feeling? An early indication of romantic intent involves lightly grasping or touching the forearm of the person of interest. Does he love me, really love me, or is he just looking to get laid? The pheromonal answer to that question is more likely to be accurate than the words that come out of his mouth. What subliminal truths do we reveal when we engage in social customs such as handshaking? It's considered high praise when we say of a man, "His shake is as good as his word." I suspect that his shake is actually much *better* than his word and is more likely to communicate an honest message.

Asexual reproduction doesn't facilitate evolutionary change in the way that sexual reproduction does. Sexual reproduction has produced a division of labor that puts the reproductive strategies of males and females at odds with each other, and out of that struggle can come some interesting innovations. There is some truth in the shorthand version of the biological difference between the sexes. Sperm is cheap. Eggs are expensive. Basic biology is at the heart of the war between the sexes. Successful female reproductive strategies focus on investing heavily in a few offspring. Successful male reproductive strategies can theoretically be judged by sheer numbers of offspring. However, the prolonged infant dependency in humans means that males and females who cooperate with each other in child-rearing have the best chance of having their own genes survive into the next reproducing generation.

Once there is selection pressure for paternal investment, males become concerned with making sure the children in whom they invest are actually their own. Concealed ovulation and continuous receptivity make certainty more difficult. After all, biologically speaking, since women invest heavily in a limited number of offspring, they want quality offspring fathered by males well-positioned in the male hierarchy. Concealed ovulation and continuous receptivity make it theoretically possible for children fathered by a high-ranking male to be ascribed to a man who, falsely believing that he's the biological father, happily invests in those offspring. (I hope it's not necessary for me to point out that I'm not defending such deceptions, but am, for the sake of advancing the argument, vastly simplifying some basic evolutionary assumptions.) This behavior launches an escalating arms race of counter-deceits and counter-detections that would favor those of both sexes who are skilled in the nuances of language.

We are a species that experienced a biological mutation that enabled reproductive deceit, and in the process of accommodating this new state of affairs, we found ourselves launched on a journey that would lead to the development of complex language skills. If one accepts the speculation that men compete with each other for the scarce resource of eggs, it doesn't take much imagination to see why a man who is able to go beyond offering protection and food in the here and now will have a distinct reproductive advantage over his fellows. It's not enough to demonstrate a willingness to invest today, in the immediate present. The reproductive success following investment that promises to continue into tomorrow and forever after (till death do us part) offers a very strong selection pressure for the future tense, a displacement feature characterizing true language. What we have managed to pull off and accomplish with this innovation has enabled us to inhabit and thrive in the far reaches of the planet. No wonder we think of ourselves as so unbelievably special, but some of what makes us unique introduces a level of anxiety that other animals don't suffer. We alone are aware of things that exist outside of the here and now. Skill with the past tense enables us to comprehend the reality of a world that existed before our individual, ego-defined selves were born, a time even before the arrival of a strange ape that later christened itself *Homo sapiens* and set about shaping the planet to its own ends. We alone are forced to face the fact that the world can and does go on without us, both as discrete individuals and as a species. We alone understand that while we are personally here right now, there will come a time when we are not. Needing to find ways to deal with the anxiety engendered by this recognition of our own mortality, we select for a worldview that solves our existential angst by developing philosophical and religious explanations of the meaning of life. Depending on your view of those two enterprises, our fascination with them could serve to bring us either closer to some objective truth or enable us to hide from it.

Chapter 9: Can We Rewrite the Script?

At my dissertation defense, one of my committee members playfully accused me of being a traitor to my sex. I suspect he was referring to my exposing the amount of power concealed ovulation gives women to direct the mating game. He apparently believed that I was telling secrets women wouldn't want to be told. In point of fact, most women aren't even aware that they're in charge. As I hope I've made clear in earlier chapters, the evolutionary directives that push us to reproduce are most effective when they fly beneath the conscious radar of both men and women. We're best at having babies when we're not fully aware of the traps we've set for ourselves and each other. Concealed ovulation and continuous sexual receptivity are first order changes, a kind of biological deception resulting from a dramatic mutation that fell outside the more usual selection process. Any subsequent deceptions employed by women were in service of the survival of the next generation of human beings. The potential costs of childbearing and rearing are so great for women—their lives are literally on the line—that they must employ every screening device available to protect themselves and their children from harm. Women are better at lying to themselves than they are to men, and it is this skill with self-deception that makes it so important to hone an equal talent for spotting a man who would exploit a woman's desire for genuine affection to indulge his own love-'em-and-leave-'em style. Both sexes have been shaped by evolutionary forces into reading lines from a script that is not necessarily to their liking. Typecasting based on little more than our genitalia can land us a role for which we might not have voluntarily auditioned. Our problem is not so much with each other as it is with the choreographer directing this reproductive song and dance routine. I understand the impulse driving those who insist there are no biological constraints on human behavior. I also wish this were the true state of affairs. Unfortunately, I'm unable to convince myself that it is. However, I am enough of an optimist to believe that it's possible to outsmart Mother Nature. We increase our chances of successfully doing that, not by pretending that there are no correlations between our X and Y chromosomes and our behavior, but by trying to understand just what those might be and figuring a realistic way to work around them. Wishful thinking is far more likely to land us in trouble than is an honest confrontation with our biological predispositions. One of the greatest errors of this sort in the last fifty or sixty years was the well-intentioned belief espoused by feminists of my generation that there were no innate behavioral sex differences between men and women. With the best interests of their sisters at heart, they further declared that there was no reason why women shouldn't approach sexual encounters with the same carefree sense of fun they assumed—probably falsely—that men enjoyed.

In 2018, a woman anonymously known only as Grace allowed her description of a disastrous date with a popular standup comedian to be published at babe.net, an online platform that has billed itself as being "for girls who don't give a fuck." Until the dust-up between Anonymous Grace and author, actor, and standup comedian Aziz Ansari hit the national print media, I'd been only peripherally aware of this drama, even though I'm constantly on the lookout for stories that attempt to shed light on the complicated nature of sexual encounters and self-described feminist Ansari had written a book on the very subject. I became inspired to look up Grace's account of her date with Ansari upon hearing that television host and fellow comedian Samantha Bee had, in a routine of her own, suggested to Ansari, "If you say you're a feminist, then fuck like a feminist."

I've now read both Ansari's book and Grace's account of her "date" and can only conclude that there are some serious misunderstandings about what constitutes an enlightened approach to sex in the 21st century. All the research I've done on the subject convinces me that he and Grace *were* fucking like feminists. That's *why* their encounter ended so badly for all concerned. Feminists are right on the money when they demand equal pay for equal work, equal opportunity in sports and education, an equal right to vote, run for office, and own property. They've done their homework when they marshal compelling data to refute the scientifically false claim that women have inferior skills in math, science, and a host of other fields. We should express nothing less than howling outrage over the failure to *hear* women's voices in all spheres.

However, when it comes to negotiating the complicated terrain of sexual communication, we've gotten an awful lot wrong. I can trace the roots of the problem back to the early days of the sexual revolution. It was the sixties, a time of rebellion against authority of all kinds, but we rarely refer to it as a social or political revolution. We talk about a *sexual* revolution. Freedom from unwanted pregnancies in the form of the birth control pill frames much of the analysis of that era, and it's true that when women have the power to control their reproductive destinies, all kinds of opportunities suddenly become available to them. However, it's also true that Mother Nature has a way of insisting on the last word.

When the pill enabled humans to divorce sex from reproduction, we had the hubris to assume we'd likewise divorced sex from its evolutionary history.

Feminists of all sexes became science deniers. They ignored the fact that *evolution is all about reproduction* and *reproduction is all about sex.* Once sexual reproduction—as opposed to asexual reproduction—evolved, male/female physical and behavioral sex differences also evolved. Telling women that as long as they didn't have to worry about unintended pregnancies, they could mimic male sexual behavior without emotional consequences was, and is, an unfortunate myth perpetrated by both sexes in a misguided quest for equality. Wishing it were so doesn't make it so, any more than a woman wishing for a penis so that she can pee more easily at a picnic enables her to magically sprout one. That said, whenever one is talking about biologically based differences—either physical or behavioral—it is important to remember the adage impressed on me in the early days of sociobiology by Lionel Tiger, my mentor and advisor at Rutgers. "Biology is not destiny; it's statistical probability."

Biological evolution moves at a glacial pace. Cultural evolution shape-shifts its way in a mad dash along the high-speed internet. It is the differential rate between these two engines of change that causes so much confusion. In the January 14, 2018, article ("The Humiliation of Aziz Ansari") in *The Atlantic*, Caitlin Flanagan says, "Sexual mores in the West have changed so rapidly over the past hundred years that by the time you reach 50, intimate accounts of commonplace sexual events of the young seem like science fiction: You understand the vocabulary and the sentence structure, but all of the events take place in outer space." She later describes the sexual mores of the 1950s, some of which persisted into Flanagan's own teenage years. "But in one essential aspect [those rules] reminded us that we were strong in a way that so many modern girls are weak. They told us over and over again that if a man tried to push you into anything you didn't want, even just a kiss, you told him flat out you weren't doing it. If he kept going, you got away from him. You were always to have 'mad money' with you: cab fare in case he got 'fresh' and then refused to drive you home. They told you to slap him if you had to; they told you to get out of the car and start wailing if you had to. They told you to do whatever it took to stop him from using your body in any way you didn't want, and under no circumstances to go down without a fight. In so many ways, compared with today's young women, we were weak; we were being prepared for being wives and mothers, not occupants of the C-Suite. But as far as getting away from a man who was trying to pressure us into sex we didn't want, we were strong."

Of course, a major problem with those old guidelines was, as Flanagan concedes, that they focused on setting women up for lives as wives and mothers. When it comes to female sexual behavior, the cultural rules are usually written by the societal demands of those in power. Fear of pregnancy was a powerful enforcer of a guilt-ridden approach to sex, as was a male-centric preference for virginal brides. The distance between the values of the fifties and today's "tindered" state of affairs is a perfect example of the rapid rate of cultural evolution. In neither of these examples have the culturally imposed sexual rules been written by the women whose lives they impacted. The hookup culture that passes as sexual freedom for today's women serves the sexual desires of a male-dominated culture more often than it gives women what they want. This is not to imply that a woman's capacity and desire for sexual pleasure is any less than a man's. There is no shame in seeking erotic gratification. Women have as every bit *as much right* to enjoy sexual activity as men do, but in order for these exchanges with men to leave us fulfilled rather than crying in an Uber, we have to start by being honest—with ourselves and with our partners—about what we want.

Anonymous Grace has taken a lot of heat. Aziz Ansari has taken even more. Grace's very insistence on anonymity would suggest that she has at least a bit of regret over her own part in the debacle. When they returned to Ansari's apartment after dinner, her deeply unsettling descriptions of his crude and rushed approach to sex made me wonder why she, who had demonstrated quite a bit of skill at getting what she wanted at the party where she first met Ansari, was willing to put up with any of it for even a single minute. Where was her voice? What in the world made Grace, who admits to choosing her outfit with such care, go along with removing it in such haste? I fault a hookup culture that somehow convinced young women that mimicking the worst components of male sexual desire demonstrates their own liberation from the sexual constraints that paternalistic systems have traditionally imposed on women. Ansari and Grace were fucking like feminists in their pretense that men and women approach sex in the same way. Except, of course, that they never got to the actual fucking. Because, of course, men and women don't really approach sex in the same way, and fucking like feminists can lead to some pretty miserable sexual experiences.

When it comes to sex, women have a lot more power and control than they recognize and acknowledge. Back in the mid-seventies, when I was recording all those hundreds of hours of interviews in which I asked heterosexual women about their sexual and romantic partners, I learned that even though the so-called sexual revolution was well underway, many of my subjects bemoaned the fact that they were at the mercy of a culture that required them to sit and wait for the phone to ring. And, on the surface, that seemed to be true. However, when I later transcribed and coded those tapes, it became apparent that women controlled a whole lot more than they thought they did, and it wasn't just through veto power.

The way in which Grace approached Ansari at the post-awards party shows her to have some skill in the exercise of that good old biological phenomenon,

Female Choice. In almost all sexually reproducing species, mate selection operates through a system in which the female of the species does the choosing, at least when it comes to sex. It's a behavioral pattern with deep roots in the economics of our evolutionary past. A woman is born with all the eggs she'll ever have. At puberty, men begin producing profligate numbers of sperm, averaging between 80 million and 100 million per ejaculation! Biological basics dictate that women be much more selective than men in their choice of partners. They have a lot to lose and must be careful not to throw away reproductive opportunities by selecting the wrong man. This evolutionarily driven behavior tends to operate whether or not women consciously want children.

Women aren't dummies, but neither is evolution. We may be programmed to reproduce, but sometimes we need to be encouraged to take those first steps by being told it's okay to fly now, pay later. Who wouldn't want the endorphin flood that accompanies falling in love? Nothing makes us feel as in-the-moment *fully alive* as sex. The old expression "love is blind" has some basis in reality. Mother Nature hopes all women will get pregnant. Most women hope the men they choose have genes that will give their children a good chance of surviving. Not only that, but most women also hope those same men will care enough to stick around and help raise the kids. This quest for good genes means that Alpha males capture our attention. However, since we're a species best described—to borrow the words of Dan Savage—as "monogamish," not every woman is going to enjoy happily-ever-afters with the guy at the top.

What's the evolutionary architect to do? Make women relatively selective. Endow women with an array of screening devices, but don't make them so fussy that they refuse to mate at all. It's when women are having trouble finding appropriate partners that they court trouble by "inventing" men. The more powerful the man, the more likely women are to move beyond passively ignoring red flags to actively ascribing traits to men that they don't actually possess.

The exercise of Female Choice flies mostly below our conscious awareness. It's easy to walk away from something we didn't realize we had. That's why we've allowed the hookup culture to rob us of far more power than it has given us. So many of our courtship behaviors happen almost on automatic pilot. Let's get back to Grace. Whether she realized it or not, she exercised a lot of agency at the start of the evening where she first met Ansari. She somehow managed to get herself invited to an event filled with powerful men. Since Grace mentioned that she showed up in a gown, I assume she selected it, hoping to catch someone's eye. It's not apparent whether Grace knew Ansari was going to be attending the post-awards party or if she identified him as a potential flirtation once she arrived, but she did what all women skilled in the unconscious operation of Female Choice do; she moved into his territory. By her own telling, Ansari brushed her off at first. At some level, Grace understood that before men find the courage to interact with a woman, most of them need to be given some pretty strong messages that the woman is not going to turn them down. (It's the *pathological* violation of these rules of consent that make the Harvey Weinsteins of the world so dangerous.) Grace hung in there. She employed their mutual affection for the old model camera they were both using to engage Ansari in conversation and some playful picture taking. Then—after she'd wandered back to the date who'd brought her there—they exchanged meaningful glances across the dance floor. In the arms of another but thinking only of you. Wow, talk about a powerful statement of desire!

This is classic Female Choice in operation. Identify a man you'd like to know better and find a way to get yourself into his range of vision. Employ signal exchange—both verbal and non-verbal—to indicate your own interest and take the measure of his. Flirtation is an important way for a woman to learn more about the object of her focus. Once Grace used these initial screening devices and identified a man she found attractive, she employed the all-important strategy of not leaving the scene of the encounter without giving him a way back into her life. Some women do this unconsciously, but it's a behavior that, once identified, is easily learned. However, Grace acquired her skill, she and Ansari had plugged each other's numbers into their phones before saying goodnight.

A couple of weeks of texting behavior followed. Grace alerted her friends to this encouraging flirtation and admitted to reading excerpts from Ansari's book, *Modern Romance*. She was doing the information gathering that is an important courtship screening device. She doesn't appear to have been approaching her upcoming date in search of nothing more than a one-night hookup. She seemed interested in making a deeper connection. Ansari made it easy for her to ascribe traits to him that he may not actually possess. The phrases "soul mate" and "true love" appear in his book so often that any young woman could be forgiven for thinking that was what he hoped to find. I wish Grace had read more than just excerpts. By the time I was two hundred pages in, I had become convinced that Ansari was more into sensation-seeking than mate-seeking. At one point, he consulted with Jonathan Haidt about the feasibility of exploiting the one-to-two-year passion peak that occurs in the early stages of a romance by seeking thrills via a series of relatively short-term relationships. In what is the best advice in the book, Haidt countered Ansari's mistaken impression regarding the rewards of such an approach and told him, "If you think the whole point of life is to gaze into your lover's eyes until you die–well, then, I wouldn't want your life."

I indulge in this bit of Monday morning quarterbacking in an attempt to illuminate what went so sadly awry between those heady moments of initial interest and the subsequent date that caused Grace to text a friend, "I hate men. I had to say no a lot. He wanted sex. He just wanted to get me drunk and then fuck me." Despite the publication of this account of a sexual train wreck on a site that claims to cater to "girls who don't give a fuck," Grace comes across as a young woman who very decidedly *does* give a fuck. She seems to understand that there are differences between the ways in which men and women approach sex. Toward the end of the evening, she told Ansari, "You guys are all the same, you guys are all the fucking same." Grace has received some criticism for feeling a bit piqued with Ansari when he failed to ask whether she wanted red or white wine. I'm less harsh in my analysis. Flirtation is a way of gathering tiny bits of information that can be individually unimportant but are still able to reveal something about a person's character. They assume significance only when they become part of a larger pattern. Female Choice enables women to assess men for traits that would make them caring partners. If Ansari's subsequent behavior had contradicted this somewhat careless attitude regarding Grace's preferences in wine, I doubt that she would have ascribed any importance to it. But his subsequent behavior reinforced her earlier impression.

I don't fault either Grace or Ansari for this date gone bad. I fault our cultural willingness to ignore the pesky biological sex differences that drive men and women to employ different mating strategies. Human reproductive success is not as dependent on playing the numbers game as Darwin would have us believe, but if even a bit of a numbers-based arms race between male and female tactics is the natural outcome of our evolutionary architecture, then a woman's one-egg-per-month option wouldn't seem to stand much of a chance against a man's 80 million sperm per ejaculation. Women counter this imbalance by being far choosier about their potential partners than men. They tend to be pretty good at figuring out which men are at or near the top of the social and economic hierarchy. This focus on selecting high-power, successful men explains the allure of celebrity. These men are surrounded by so many willing women that the worst of them may tend to treat women as disposable objects. A woman who heads straight for the alpha male should be aware that the odds aren't in her favor. Nonetheless, the behavior is understandable. Grace was pretty savvy about assessing Ansari's potential as a romantic partner. Had she been one of the women I interviewed, I would have taken note of the fact that while eating dinner in a riverside restaurant, she had been sufficiently impressed to pause and text a photo of her meal to friends who were waiting to hear how it all was going. Courtship screening behavior in the age of social media. Although troubled just a bit by Grace's willingness to sacrifice personal intimacy in favor of public display, I would have coded the attention to the food Ansari was providing as evidence of the erotic allure of courtship feeding behavior. Her subsequent mention, in the story on babe.net, not only of Ansari's exclusive address but also of the other famous residents who lived in the building would have prompted me to check the box indicating her awareness of Ansari's control of material resources as a factor influencing her attraction to him. I don't intend for my objective analysis to take the romance out of what was, at least up until the point when the pair returned to Ansari's apartment, a rather encouraging reinforcement of Grace's growing feelings for Ansari. That said, it doesn't hurt women to be at least somewhat aware of factors such as wealth and celebrity that can blind them to what they might otherwise identify as red flags or at least yellow ones urging caution. Ansari's hasty rush to leave the restaurant and head back to his place might have put Grace on alert had she not already fallen prey to the power of his name recognition.

Many years ago, Robin Fox, one of my mentors at Rutgers, countered the assumed rigidity of viewing human behavior through a sociobiological lens with a bit of wisdom from Katharine Hepburn in her role as Rose Sayer in *The African Queen.* The Humphrey Bogart character, Charlie Allnut, defends behavior that Rose finds objectionable by informing her that it is only human nature. She counters by telling him, "Nature, Mr. Allnut, is what we are put in this world to rise above."

However, before we can rise above it, we have first to identify it. It's difficult to outsmart behavior that interferes with us having a reasonable measure of control of our lives by pretending that it doesn't exist. Men and women approach sex differently. In knowing how to prevent pregnancy, we've successfully risen above nature. We've had a more difficult time rising above the *behaviors* that reinforce the cheap sperm, expensive egg biological basics. Those annoying whispers directing women to be selective and telling men to wait for encouragement before making sexually overt moves are difficult to silence. And perhaps we'd be better off if we stopped trying so hard to ignore wisdom that has been sculpted by millions of years of evolutionary history. The female of the species is designed to be in control of reproductive decisions. Women should not relinquish the power that is their birthright. The fact that neither men nor women seem able to function in today's hookup culture without prodigious amounts of alcohol is a warning sign that we are messing with Mother Nature in ways that are unlikely to make anyone happy. If women have to be drunk out of their minds before enduring hookup sex, something is seriously wrong. If it takes keg parties or hours at a bar before men can overcome their fear of rejection, then neither sex is honing their skills at practicing the exquisite and complicated system of signal exchange that characterizes human courtship. Unless and until we learn to decipher the messages we are sending each other, neither men nor women are likely to have experiences that lead to the emotionally satisfying relationships that both claim to want.

Chapter 10: Designing Women

If modern women are drawn to men who offer evidence of their wealth and power, I suspect it is because women who selected for those traits in our evolutionary past were more likely to have given birth to babies that survived and grew into reproductively successful adults. If today's *Homo sapiens* are technologically sophisticated, I suspect it is because early human females thought males who'd figured out how to invent simple tools that enabled them to exploit otherwise scarce resources in a harsh environment were quite sexy indeed. Do humans have a way with words that isn't found elsewhere in the animal kingdom? Blame the seductive charm of sweet nothings that weren't really about nothing so much as they were about promising to love, protect, provide, and nurture. By selecting for capabilities that optimized survival in our evolutionary past, women have been responsible for endowing modern humans with the traits that make us who we are.

Evolution works exclusively in the here and now. It does not have a future tense, but this new species of ours and its emerging skill with language was well on the way to figuring out how to add that particular point of view to its other talents. Human females became so adept at appraising how a given male's knack for manipulating the environment might translate into security for her and her offspring that, in a very real sense, they were engaged in *designing* traits that would be expressed in the next generation. Early women who were good at anticipating what the future might hold became the engineers of models who were almost pre-adapted to that future. And what happened if they guessed wrong? Natural selection has a way of eliminating our mistakes. Our initial human reproductive innovations made it possible for females to engage in a trial-and-error exercise in intelligent design. They suddenly had real power to choose their sexual partners and, unlike chimps, weren't most often impregnated by a male whose predominant skill was simply bullying his way through the crowd of other males and their competing sperm. Early human females who selected for reproductive partners with technological skills were moving the species in a necessary direction. After all, Prototype #1 was poorly equipped to make it without some kind of clothing with which to garb his exposed skin, shelter to protect him from the elements and predators, tools to enable him to exploit resources neglected by his nearest primate cousins, and weapons to make up for his lack of fangs and claws. The development of all these things was expedited by the development of language. The subtitle of Terrence Deacon's book says it all. Language and the brain *co-evolved.*

At this point, it may be important to recall the widowbird and his recklessly marvelous excess of tail feathers. Bowerbirds. Manakins. Resplendent Quetzals. Grassfinches. Platyfish. Túngara frogs. If Michael Ryan's premise about the seductive power of *more* is as true as I believe it to be, human females have taken the game to a whole new level. Early humans needed to figure out how to compensate for their vulnerability to wind, rain, snow, heat, cold, and predators. In addition, we needed to find and consume enough calories to fuel our muscles and energy-hungry brains, but why were we unable to stop once we had achieved a measure of protection sufficient to keep us alive and reproducing? Why is it not enough to have a roof over our heads without aspiring to a mansion with a Beverly Hills address? Does our food really have to be plated into works of art served in exquisitely designed restaurants by hushed waiters schooled as to the provenance of every single ingredient right down to the salt? Clothes that once served no purpose other than protection are now trumped by those with designer labels, even though the high-end styles may be far inferior in terms of keeping us warm and dry should we encounter a downpour or sudden drop in temperature. Examine any aspect of human life, and it becomes apparent that enough is not now and never has been *enough*. We impose our infinite desires on a finite world. We love our things, our material possessions, and burden ourselves with them and the need to find places in which to store and display them. We are in a state of denial about the damage all this consumption causes. I'm always struck by TV coverage that follows the increasingly common weather-driven natural disasters. Homes are buried by mudslides, leveled by out-of-control fires, submerged under rising waters, and flattened by tornadoes, and when a family lucky enough to have gotten out in time is interviewed by the eager newscasters, they invariably say, "At least we all survived. Our belongings can be replaced." They hug their kids. "Everything that really matters is right here." And it's true. We do cherish human connections more than we care about material possessions. Firefighters risk their lives to save the family dog. We rightly declare them heroic and celebrate the values they reflect. We are good at seeing truth in the small here-and-now scale and excellent at denying evidence of it in the larger picture. Because here's a simple fact we aren't willing to face: by refusing to get serious about addressing something as existentially threatening as climate change, we as a species are demonstrating that, in the long run, we value our immediate material plenty more than we care about our children, grandchildren, and the future we are handing off to them. I don't intend to imply that the families impacted by these tragic climate events could have sidestepped the catastrophe if only they'd placed fewer orders with Amazon, avoided air travel, recycled their plastic bottles more efficiently, or perhaps avoided plastic altogether. Our overconsumption is a collective phenomenon. It has to be dealt with on a grand scale. That's not to say that the small decisions each of us makes on a daily basis don't matter. They do. Individual actions *are* important, but nowhere are they more important than when we demand sweeping policy changes from our elected officials and refuse to elect those who have failed to educate themselves about the urgency of the problem. Effective legislative solutions can incentivize a sustainable approach to agriculture, manufacture, and consumption so that the costs and benefits are fairly distributed across all levels of society. In the final chapters, I'll discuss what those legislative solutions might entail, but for now, let's get back to Darwin and the ways in which flaws in some of his early theories may have set the stage for the fix in which we currently find ourselves.

Darwin's original sense of fitness simply meant that an animal was well adapted to its environment. If we were able to take a truly objective measure of our biological fitness, would we be quite as willing to claim human superiority in all things? When it comes to comparing ourselves to other creatures, we've fallen into the trap of assuming that *different* means *better*. We have such a narcissist's love for our own reflection in the gene pool that we are blinded to the downside that accompanies some of what makes us unique. How can anyone seriously claim that an animal that is destroying the environment on which its very existence depends is demonstrating Darwinian fitness?

I have a colleague who once told me about an ongoing argument he'd been having with an associate who was convinced of the superiority of human intelligence. My friend had taken the position that dolphins were likely to be every bit as smart as humans. His companion in debate finally came up with what he thought would be the last word in their exchange. "Dolphins can't build skyscrapers." Given my own skepticism about the superiority of all things human, it's no surprise that I think my friend's response— "They don't have to."—settled things once and for all.

In fairness to Darwin, he *was* worried that humans might be doing something wrong. Unfortunately, he wasn't worried about either excess consumption or overpopulation as much as he was concerned about the reproductive deficiencies of those who were already doing the most consuming. His blindness to the impact his favorite species was having on the natural world is understandable. In the mere one hundred and fifty years since *The Descent of Man and Selection in Relation to Sex* was published, the runaway pace of material innovation has changed the face of the planet in ways that only science fiction writers may have foreseen. Employing the cleverness that is our claim to fame, humans have turned the evolutionary process on its head. Prototype #1 was born deficient in enough areas that we had to pretty quickly figure out how to reconfigure the world to our own ends. We didn't have time to wait until natural selection's more patient process had fashioned *us* to fit *it*. And now the forces of cultural and technological evolution have altered the environment in which we must survive more rapidly than biological evolution can adapt, so fast that not even the inventions we design to address the problems seem able to keep up with the rate of change. I sometimes think of life on earth as a single organism and human activity as the rapidly growing cancer that threatens to snuff it out. In our arrogance and exclusive focus on ourselves, we forgot to pay attention to the interconnected web of life through which every living thing is somehow dependent on all the others. Here's a hard truth: it turns out that we need the honeybees far more than they need us. We can only hope that species with shorter generational spans than those possessed by humans manage to adapt more quickly than we can. But, unfortunately, when it comes to the effects of climate change, I'm not picking up enough documentation of adaptations in creatures, either large or small, to inspire much optimism. The dramatic alterations in global ecosystems that define the past 40 years serve as a natural experiment in evolution. We see an increase in extinction events as well as at least temporary survival by migration, but I'd feel better about the adaptive power of both sexual and natural selection if evidence substantiating beneficial evolutionary changes in at least some species began to make the scientific headlines.

Darwin's primary concern was not that humans were destroying the environment but that the way in which we were playing the reproductive game would, in his words, cause the species, *our* species, to "retrograde". Darwin's understanding of how natural selection worked—an understanding he claimed was inspired by Malthus's *An Essay on the Principle of Population*—was focused squarely on the numbers game.

> This is the doctrine of Malthus, applied to the whole animal and vegetable kingdom. As many more individuals of each species are born than can possibly survive; and as, consequently there is a frequently recurring struggle for existence, it follows that any being, if it vary however slightly in any manner profitable to itself, under the complex and sometimes varying conditions of life, will have a better chance of surviving and thus be naturally selected.

Malthus's stated fear was that the rapid growth of human populations would outstrip the ability of the food supply to keep pace. Although his theory has become so widely accepted as to have entered the realm of "everybody-knows" common knowledge, Malthus himself never saw the need to document his treatise with actual measurements. In the introductory paragraph, he apologized that the rush into print prevented him from providing facts. However, the lack of empirical support for his statement appears not to have shaken his confidence in its truth. As Malthus explains, referring to himself in the third person:

> He presumes, however, that the facts which he has adduced will be found to form no inconsiderable evidence for the truth of his opinion respecting the future improvement of mankind. As the Author contemplates this opinion at present, little more appears to be necessary than a plain statement, in addition to the most cursory view of society, to establish it.

Malthus' argument that population size increases exponentially while the means to feed everybody increases only geometrically was primarily grounded in a concern about the rapid growth in those segments of the population that *he* deemed undesirable. Malthus bent his theory in a direction that would support his bias against certain *kinds* of people. That it continues to masquerade as science is perhaps best explained by the human penchant for tribalism. I don't know how else to explain the willingness to find logic in his assumption that any exponential population growth exhibited by humans with our long generational spans and relatively low birthrates would fail to be matched by a corresponding exponential increase in the reproductive exceptionalism of the rabbits on which we fed. If Malthus had sketched his argument in terms of the finite limits of the earth against the infinite nature of human desire, he could have made a stronger case, but then, of course, he would have had to abandon his prejudice against the poor and the Irish.

However Darwin actually came to his understanding of the way in which species evolved, he claimed that a Malthusian flash of insight convinced him that natural selection needed large numbers of individuals in order to weed out those less able to survive. This is what biologists refer to as death rate control of population size, and it is more difficult to demonstrate than one might assume. Unable to abandon this initial Malthusian focus on numbers, Darwin then concluded that the individuals best suited to their ecological niche ought to indicate their superiority by leaving more progeny than their less fit competition. Nature red in tooth and claw. And there *is* that aspect to survival, but the violent side of nature is more apparent between species than it is within species and more common between groups than within groups. *Homo sapiens* is a social species and *cooperation* among members of a population has done more to ensure its survival than has an internal struggle. Generosity and a loose constellation of traits that I coded as "control of social resources" rank high on the list of characteristics that add luster to man's sexual aura.

The man we credit with being the father of the theory of evolution was convinced that within a species, those groups having the most offspring would outcompete their neighbors. It was the traits of the profligate reproducers that would get passed on and become enshrined by the forces of natural selection. If his argument seems just a bit slippery, it's because it was. That tends to happen when objective observations of birth rates are contaminated by a subjective take on what constitutes fitness. In the years since Darwin published his seminal work, our understanding of evolution has gone from focusing on group selection to considering the relative success of individual actors, and finally, due to work popularized by Richard Dawkins in his book *The Selfish Gene*, to viewing selection pressure operating at the level of the gene as the path offering the most accurate analysis of what changes are actually adaptive. Gene frequency is considered a reliable measure of reproductive success. Remember the deep-sea-diving Bajau? The frequency of alleles correlated with that proficiency can be viewed as a measure of the reproductive success of a genetic pattern that is unique to that population. Understanding whether those changes were driven by sexual selection, natural selection, or both is not my focus at the moment. Coming up with a working definition of reproductive success is. The increase in the number of genes associated with Bajau diving ability is a measure of the reproductive success of those genes. They had proven beneficial enough to be successfully replicated in the next generation.

Darwin may not have known about measuring gene frequencies, but it was fairly easy to count the number of *individuals* in a given segment of a population, particularly when the researcher himself belonged to the group being counted. He was not encouraged by the conclusions forced on him by the available demographic evidence. Poor Darwin, everywhere he looked he found reason to believe that the reproductive behavior of the species nearest and dearest to his heart would—unless changed—reverse the progress that had enabled it to rise above primeval man, his apelike progenitors, and the lower animals. Darwin is obviously troubled as he summarizes, in *The Descent of Man*, the findings of his own research and that of some of his contemporaries.

> A most important obstacle in civilized countries to an increase in the number of men of a superior class has been ... the fact that the very poor and reckless, who are often degraded by vice, almost invariably marry early, while the careful and frugal, who are otherwise virtuous, marry late in life so that they will be able to support themselves and their children in comfort. Those who marry early produce within a given period not only a greater number of generations but, as shown by Dr. Duncan, they produce many more children.

> The children, moreover, that are born to mothers during the prime of life are heavier and larger and therefore probably more vigorous than those born at other periods. Thus the reckless, degraded, and often more vicious members of a society, tend to increase at a quicker rate than the provident and generally more virtuous members. Or as Mr. Greg puts the case: 'The careless, squalid, unaspiring Irishman multiplies like rabbits; the frugal, foreseeing, self-respecting, ambitious Scot, stern in his morality, sagacious and disciplined in his intelligence, passes his best years in struggle and in celibacy, marries late and leaves few behind him. Given a land originally peopled by a thousand Saxons and a thousand Celts–and in a dozen generations, five sixths of the population would be Celts, but five sixths of the property, of the power, of the intellect, would belong to the one sixth of the Saxons that remained. In the eternal struggle for existence, it would be the inferior and less favored race that had prevailed–and prevailed not by virtue of its good qualities but of its faults.'

Of course, Darwin is making a logical error here, one that continues to be made by those who, like Darwin before them, enjoy lives of privilege. Confusing wealth with genetic superiority is a mistake made most often by those with great fortunes. When Darwin talks about "men of a superior class," he is relying on a cultural measure of superiority, not a biological one. In the early pages of the chapter containing the above paragraph, Darwin references Alfred Russel Wallace's understanding of the unique nature of human adaptation.

> Mr. Wallace, in an admirable paper before referred to, argues that man, after he had partially acquired those intellectual and moral faculties which distinguish him from the lower animals, would have been but little liable to bodily modifications through natural selection or any other means. For man is enabled through his mental faculties "to keep with an unchanged body in harmony with a changing universe." He has great power of adapting his habits to new conditions of life. He invents weapons, tools, and various strategies to procure food and to defend himself. When he migrates into a colder climate, he uses clothes, builds sheds, and makes fires; and by the aid of fire cooks food otherwise indigestible. He aids his fellow-man in many ways, and anticipates future events. Even at a remote period, he practices some division of labor.

Wallace goes on to point out that when the physical structures of lower animals fail to keep pace with greatly changed conditions, those animals will, unlike humans, go extinct. It wasn't only Darwin who credited Thomas

Malthus with inspiring his understanding of how natural selection worked. Wallace made the same claim. However, toward the end of Wallace's life, he rejected both the ideas of Malthus and the belief that natural selection was sufficient to explain the human tendency toward progress. While never entirely abandoning his early interest in science, Wallace turned increasingly to spiritualism in his struggle to make sense of human primacy and to land nationalism in the hope of addressing inequality. If only he had taken the time to follow Darwin's arguments on sexual selection to their logical conclusion—something that not even Darwin had been able to do—he might have recognized in female choice and in the talents of those designing women a biological instrument capable of shaping human development that would have obviated the need to embrace spiritualism in order to explain the origin of human progress. However, given the era in which both Wallace and Darwin lived and worked, it's probably too much to expect that they would have been able to grant their female contemporaries a role of much significance when it came to directing the evolutionary process. That said, although Wallace continued to distance himself from Darwin's theory of sexual selection, his lifelong quest for social and economic justice eventually led him to propose that the economically enabled and enlightened socialist women of the future would bring *Homo sapiens* back under the sway of natural law and advance human progress when these women chose only men who met their exacting standards. Wallace made a point of noting that those standards were unrelated to non-utilitarian notions of beauty. Because Wallace fell into the old trap of thinking that sexual selection resulted only in aesthetic traits without functional value, he was able to hold firm to his rejection of Darwin. By stressing the *range* of positive male attributes that would influence the choices exercised by feminist women of the future, Wallace apparently believed that his position was intellectually consistent with his repudiation of Darwin's theory of sexual selection. In other words, if the traits that influenced female choice weren't sexually dimorphic, they couldn't possibly fall under the umbrella of sexual selection. In point of fact, the system he envisioned was a much more *accurate* description of the way female choice and sexual selection actually operate than the one that Darwin had advanced.

Wallace may have believed that the kinds of enlightened reproductive decisions that he thought women were capable of employing had to wait for some utopian future, but I feel certain that women have always had far more say in choosing their partners than either the males or the females engaged in the human courtship dance have ever acknowledged. Women—and females of all species—were then and are now nature's "invisible gorillas." It's easy to miss what we are doing. However, in non-human species, that all changes when a female is in estrus. For the brief period that a female is "in heat," she has the full and focused attention of every male in her troop, pack, herd, or neighborhood. She doesn't even have to be in their immediate vicinity; her powerful pheromonal message will communicate the existence of her available eggs more effectively than any profile on Tinder. Males are very good at noticing a female who is flaunting her fertility, but once that phase of her reproductive cycle passes, she tends to fade back into relative obscurity. She's given a little breathing room and returns to the hard work of making sure her offspring receive the care and sustenance needed for survival. Once a female is no longer sexually available, the males tend to ignore her and, once again, devote their attention to the testosterone-infused fellows with whom they compete for positions in the hierarchy.

The human case is a very different story. Concealed ovulation is almost certainly a contributing factor when it comes to how very difficult it is to *see* women's behavior. Our reproductive invisibility makes it possible for us to learn a lot about the men in whom we think we might be interested. Sexually speaking, we don't necessarily *want* a man to notice us, at least not at first, not before we determine whether or not we're in the market for the kind of attention he might be "selling." However well that may work for us reproductively, it works against us when this talent for ignoring women is carried over into the classroom, the workplace, and our professional lives. Female contributions are overlooked, undervalued, and attributed to male peers. Women resist being seen primarily as sex objects and rightfully resent being described in terms of appearance and the clothes they wear. Much has been made of the role of fashion in human courtship. We're damned if we do and damned if we don't. I've heard people puzzle over why it is that male birds are the flamboyantly colored sex, while modern humans seem to have reversed the roles. As previously mentioned, the birds with gorgeous feathers are mostly advertising maleness. I suspect the same is true of female fashion. Lacking a visible estrus cycle, it's one way of reminding the boys that we're girls. Men do pay more attention to pretty women, but they are criticized if the manner in which they dress makes the message too sexually overt. The cover of Evelleen Richards' book, *Darwin and the Making of Sexual Selection*, features an illustration of a fashionably dressed Victorian woman wearing a peacock on her head, its gorgeous tail feathers trailing down her back like a cape. Even her parasol features those beautifully iridescent eyespots. Lacking a pronounced breeding season or attention-getting estrus cycle, it's almost as if human females are forced to employ fashion as a way of making sure men know they're in the room. Even when seen, it is far too easy to begin *and end* by focusing on how women dress and their physical appearance. But we shouldn't forget that those big brains of ours are located equally within male and female skulls. Men and women are both far too good at ignoring the importance of female intellectual contributions and depriving themselves of the problem-solving abilities of more than half the population. Historically, we've turned our backs on an awful lot of brainpower. That said, hard though it may be to rise above biological dictates, it can be done, and we are doing it. The opportunities open to me, my daughter, and my granddaughters are dramatically more varied than those available to Emma Darwin and the women of her generation.

Although neither Charles Darwin nor Alfred Russel Wallace could have anticipated the changes in the traditional roles of men and women in the twenty-first century, I suspect that Wallace would have been the more likely of the two to view them favorably. Wallace, a land surveyor by trade, had a workingman's appreciation for how difficult simply procuring the means of subsistence was for many people. He was concerned about the inequality made possible by extreme concentrations of wealth and understood that much of the disparity had its roots in land monopoly. Wallace was sufficiently impressed by the theories of Henry George, an American economist, that he wrote to Darwin recommending that he purchase a copy of *Progress and Poverty*, George's seminal work, in which he proposed addressing growing inequality and the boom/bust nature of the business cycle by eliminating all taxes except one imposed on land values. George used the term "land" to refer to *all* naturally occurring resources, a point that has been glossed over by his critics but one that merits a second look as we search for policies we can employ to impose restraint on our wanton destruction of the planet. In the mid-1800s, when both Alfred Russel Wallace and Henry George were searching for ways to guarantee that all humans might have access to the means of existence, land was the most economically obvious natural resource. It's understandable that those hoping to dismiss George in 2021 would claim that while land was central to the economy of agriculturally based societies, his work could have no significance in modern times. His critics conveniently overlooked George's broader use of the word "land," one that included pure water, clean air, virgin forests, mineral reserves, ocean fishes, and, yes, the land under our feet.

Monopoly in naturally occurring resources of all kinds is the root cause of the current worldwide growth in economic inequality and is a major player in the degradation of the environment. Monopoly in land was the primary factor responsible for Irish poverty in the 19th century. In the next chapter, I'll explore the relationship between that poverty and the high birth rates that troubled both Darwin and Mr. Greg, the correspondent whose notes on population Darwin enthusiastically referenced when writing *The Descent*. Mr. Greg correctly anticipated the future disparity in land ownership between the Celts and the Saxons, but he reversed cause and effect when he blamed the outcome that so worried him on the "careless, squalid, unaspiring Irishman" who lacked the restraint that enabled "the ambitious Scot [to pass] his best years in ... celibacy."

"Given a land originally peopled by a thousand Saxons and a thousand Celts–and in a dozen generations five sixths of the population would be Celts, but five sixths of the property, of the power, of the intellect, would belong to the one sixth of the Saxons that remained." Wow. The above description does seem to predict some of the issues raised in 2011 when the Occupy Wall Street movement protested against the concentration of wealth by the top one percent. However, Mr. Darwin and Mr. Greg, these gentlemen of the propertied elite, seemed far less concerned about the monopoly holdings of the "stern-in-their-morality" Scots than they were about the sexual behavior of the poor and "multiplying-like-rabbits" Irish. There *is* something unusual about human reproductive behavior but, as I hope to explain, the solution to the problem Darwin describes is to be found in an expanded understanding of how the changes associated with that initial chromosomal fusion made it possible for humans to introduce a dramatically different twist on the way they operate economically. Suddenly an individual *Homo sapiens*— unlike the individual members of other species—could "own" resources well in excess of those he could claim and defend using nothing other than his physical strength. Selection for symbolic skills enabled a single person to greedily control far more wealth than he or any of his descendants could ever possibly need. Not only that, but his claim to have written title to the natural resources in question meant that he had the law on his side. Something is seriously wrong when one person's ownership is allowed to prevent large numbers of his fellow humans from being able to meet even their most basic needs. Whenever we see wild variance and distortion in an economy's distributive function, we're going to witness the phenomenon that Mr. Darwin and Mr. Greg found so troubling: an increase in the birthrates of the impoverished.

Designing women select for traits necessary to survival. Since our most immediate needs are material ones, food and shelter, there is an understandable emphasis on traits reflecting a man's wealth, his ability to provide. However, that is not all there is. The best providers need to demonstrate other capabilities as well. Being able to recognize and figure out ways to exploit an underutilized resource in a particular environmental niche requires intelligence. Humans are a social species, and the most effective and ultimately most successful individuals use cooperative efforts to maximize their own rewards while at the same time enriching not only those in their face-to-face networks but also those who are geographically removed. We trade, sometimes physical goods and sometimes a talent unique to one person, for abilities in a skillset possessed by someone else. It's the rising-tide-lifts-all-boats way of approaching economics. Economic historians predict that no society can long sustain the levels of inequality we are rapidly approaching. Greed taken to extremes becomes maladaptive. How would it benefit the reproductive success of any woman if the man she chooses is driven by avarice and unwilling to give of himself and his material goods? In response to this emphasis on rearing healthy children, women select for many traits in addition to wealth and control of material resources. Intelligence. Control of social resources. Generosity. Protective behavior. Tenderness with children. Rank relative to other males. Willingness to share food.

We need to get serious about figuring out ways to stop an increasingly small minority of individuals from employing their affection for money and power with such ruthlessness that their pathologically acquisitive behavior and distorted levels of control destroy the planet's ability to provide for us all. In order to halt the destruction before it's too late, it's time to turn our attention to the other human traits that women have helped configure and learn how to use them to solve the very problems that we, ourselves, have caused. There is more to our species-specific behavior than greed. A lot more. We can reverse our destructive tendencies by drawing on our creative ones, but first we need to consider what makes our approach to our ecological niche so different from the way other animals and plants demonstrate their fitness.

Chapter 11: Painting by Numbers

Darwin was puzzled. He'd been led astray by Thomas Malthus and his emphasis on sheer numbers as expressed in Malthus's 1798 publication, *An Essay on the Principle of Population*. I don't think either Malthus—who died in 1834, twenty-five years before the publication of *The Origin of Species*—or Darwin was consciously racist, but they, like others before and since, were very concerned about the differential reproductive rates between the ethnic group with which they identified and that of those in groups they viewed less favorably. Darwin came from a family of abolitionists and—in contrast to other evolutionary biologists of his era, most notably Louis Agassiz—he had, for his time, a fairly enlightened view of "race." His record on gender, ethnicity and social class is less exemplary. It was the reproductive behavior of the Irish that both Malthus and Darwin found so troubling. They believed the Irish were simply having too many children, and Darwin thought this tendency of the poor to out-reproduce the rich would inevitably lead to the degradation of the human species.

Darwin believed that evolutionary success was directly related to numbers of offspring and the resultant struggle for existence over scarce resources. Natural selection would work on slight spontaneous variations between individuals and favor those who were best able to survive and demonstrated their superiority by having lots of progeny. It was a continuing fight over limited food and shelter that drove evolution. But if the battle over nature's insufficiency is such a central player in determining evolutionary outcomes, how is it that when we look at other species, we find far less of the struggle-for-existence sort of behavior than that exhibited by humans? Other members of the animal kingdom aren't characterized by scenarios in which certain individuals or even whole segments of a population are routinely denied access to the things they need in order to live. Consider the Dark-eyed Junco. The slate-colored "race" doesn't thrive while its pink-sided relatives starve. Nor do these birds exploit and expropriate the seed-gathering labor of the gray-headed branch of the family so that the Oregon "tribe" might hoard and stockpile food far in excess of anything it or any of its descendants might forever need. Despite our reputation as a social species, humans don't always appear to make the social good a primary focus. Where have we gone wrong? The natural world is capable of providing enough for everyone. How is it that masses of people are forced to live right at the knife's edge of subsistence, not dying, but not exactly thriving either, while others wall themselves off in gated communities and nations and enjoy lives of plenty? This doesn't look like the work of a species that coined the phrase, "the milk of human kindness." How is it that other creatures manage to share access to the food, water, and shelter necessary to life without ending up with the life-threatening inequalities characterizing human populations? They do it by defending their right to certain territories through threat displays that are usually sufficient to drive interlopers off and, should those fail, by resorting to the more costly strategy of actual combat. Not every breeding pair or population gets access to prime real estate, but even those that are forced to settle for something less manage to establish themselves on a roughly proportionate site. The weapons employed are those possessed by their physical bodies: size, strength, antlers, hooves, fangs, claws, and gestures of intimidation. Because members of the same species are pretty much matched in size and weight, so are the territories they control reasonably approximate. One doesn't have to be a biologist or ornithologist to have observed the skill with which other animals establish a fairly equitable distribution of resources. The birdsong that each spring turns our backyards into symphony halls is probably as much about demarcating borders as it is about luring the ladies. However, in birds, as in humans, it may be difficult to separate sex appeal from demonstrations of the ability to provide. Not every male mockingbird manages to find a mate in any given season, but most females do. Once breeding territories are established, the borders are respected without resorting to life or death battles. And what happens to those feathered fellows who remain bachelors? They mostly forage outside the established territories and live to try their luck in another season.

How do humans defend their turf? On a national level, it can, as the human predilection for armed conflict makes clear, get very bloody indeed. However, on a personal level, very rarely do we resort to fisticuffs. There is no need. Our symbolic skills have taken the defense of turf to a whole different level. Instead of using song to establish our claim to the resources needed to survive, we rely on laws. And laws rely on representations that have evolved beyond musical notes. The same symbolic skills that we employ to create works of literature and build suspension bridges also enable us to construct laws that proclaim our right to occupy and use certain spaces to the exclusion of others. Once property rights are enshrined in the documents governing nations, they become the most vigorously defended of those laws. This seems intuitively fair to us. After all, property laws serve to protect what, *at first glance*, we seem to have earned through our own industry and cleverness. Our laws function to keep others out much as the song of one bright red cardinal functions to tell another that I have worked for this and it is, therefore, mine.

But if we are really playing the game by the same rules that other species employ, how is it that so many of us starve and so many of them do not? Because our skill with symbols has been used in a manner that rewards the children of those who already have and discriminates against the children of those who have not. The offspring of the fortunate few use their access in such a way that they are able to perpetuate and magnify the accumulated wealth and power that they consider their birthright. They unwittingly mobilize their advantage to keep others out. No longer do we divide up the gifts of the natural world in a way such that each of us has access to the food and shelter we need in order to survive. What has happened to make this distortion possible? Enter money, a glorious innovation that keeps track of who owes what to whom, facilitates exchange and enhances the ability of people, whether in the same town or geographically removed from each other, to interact in economically productive ways. We can cooperatively pool our specialized skills and talents in the creation of goods, services, and inventions, all the things that enable us, in Wallace's words, "to keep with an unchanged body in harmony with a changing universe." But the *symbolic* representations of value employed in tracking our various exchanges of labor and material and cultural goods can also be *hoarded* and used to buy up and establish titles to and monopoly holdings of the commons in nature that will be staunchly defended in courts of law. It's through behavior such as this that our species can get itself into the kind of trouble that creates extremes of inequality not found elsewhere in the animal kingdom.

Here's the thing about the commons in nature. No human being created it with his or her own labor, and we all need the things it contains in order to live. No animal, human or otherwise, can exist without a place to physically inhabit. We are land animals and need some plot of ground beneath our feet and some form of shelter over our heads. We also need pure water, clean air, and the gifts provided by nature's bounty. Humans were hunters and gatherers long before we were agriculturalists. Some human societies still are, but just because it *can* be done doesn't mean that I'm arguing for a return to that time. I like hot running water, indoor plumbing, and other creature comforts as much as the next person. I simply want to make sure that my plenty doesn't come at someone else's expense. But before elucidating ways to prevent the monopoly of the commons in nature and other arenas, I want to return to Darwin's fear that when it came to reproductive behavior humans were doing something wrong.

It's not that we were doing something wrong; it's that we were doing something *different*. Modern biologists determine reproductive success by counting gene frequencies. Since tallying up actual genes can be a cumbersome affair, we usually accept counting numbers of individuals as an acceptable equivalent. This measure of reproductive success provides a fairly accurate assessment of which traits are adaptive and at least when it comes to measuring the economic impact—read control of resources—of the individuals carrying the genes in question; it works for all species *except humans.* I argue that we have overlooked something in our calculations of human reproductive success. We have failed to account for the differences in economic behavior between individuals. Reduced to its basics, economic behavior is simply a concept used to describe the resource accruing behavior of an individual or a species. Robert Trivers differentiated between the resource-accruing ability of one male *Anolis* lizard and another by weighing and measuring them. He found that slight differences in size influenced the reproductive choices of the females, who preferred larger males. In the case of lizards, resource-accruing ability assumes a close correspondence with the genes determining size. Reproductive success was measured by observing the greater receptivity of females to the larger males. There are biological constraints on what constitutes optimal size in lizards. Therefore, the differences between one lizard's resource-accruing ability and another's are *almost* constant and will never begin to approach the range of differences in material and social control found between individual humans. The symbolic skills exercised by *Homo sapiens* have no physiological limits and have enabled us to break the tight correlation between physical strength and reproductive success, at least as determined by counting numbers of offspring.

Given the complexity of human cultural and social institutions, when considering the factors contributing to our reproductive success, we need to take into account the distinction between resource-accruing *ability* (RAA) and resource-accruing *opportunity* (RAO). Lizards enhance the likelihood of getting their own genes in the gene pool by optimizing mating opportunities. For them, it's pretty simple. For us, it's pretty complicated. Those with solid control of wealth and power can enhance the chances of any child they have by devoting lots of material and social resources to that child's growth, development, education, and success-enabling connections with other powerful humans. Those at the top of human hierarchies don't need to have lots of kids in order to have a good chance of getting their own genes into the gene pool of future generations. Whatever small number of children they might decide to have are so cosseted and cared for that they will almost certainly reach adulthood and will, themselves, become highly sought-after reproductive partners. The higher mortality rates afflicting those at the bottom of human hierarchies mean that folks who struggle economically need to have many children in order to be certain that at least some of them will survive. Humans have mastered symbolic control of resources and no longer depend on physical size and strength to demonstrate fitness in this area. Think of it as a simple equation, one expressing rough equivalence rather than absolute equality. In all species *except humans*, reproductive success (RS) can be determined by simply counting the number of individuals or its genetic equivalent, gene frequency (GF). RS = GF. When dealing with humans, we need to add a multiple to our computations of reproductive success. If our predictions are to be accurate, we must determine not only the number of children but also the extent of the social and material resources that the parents are able to devote to each one of those kids. RS = GF x RAO.

Selection for *more* has some built-in limits in the animal kingdom. If your tail feathers get too long or your antlers too heavy, it can spell, if not extinction, at least an unacceptable level of individual risk. This imposes a sort of check on carrying the seductive quality of *more* to extremes. The female of the species can select for the super-stimulus of the trait that triggers attraction, but built-in physiological limits keep things from getting out of hand. And that's pretty much how things operate in all species except *Homo sapiens*. Because we humans exercise control of resources *symbolically*, we don't have *biological* limits on how much a single person can claim to own. The reason we can—when calculating reproductive success in other species—safely ignore differences in the resource-accruing ability between individuals is because they aren't very dramatic; the resources needed are pretty much uniform, at least as compared to the case in *Homo sapiens*. Although there are variations between one anole and another that are noticeable to the females doing the selecting, those discrepancies pale in comparison to the differences in resource-accruing ability between one human and the next. Early human females were selected for language fluency and other symbol-making abilities. It is our skill with symbols that has made possible the extreme differences between a Jeff Bezos and the guy who earns his living programming computers. Because the wealth disparities between individual men *are* dramatic, we are forced to take into account the differences in resource-accruing opportunity between one human and another when calculating reproductive success.

Think of it this way. When measuring reproductive success in humans, we can't be content with simply counting numbers of humans. If Darwin had wanted to reassure himself with an accurate comparison of the evolutionary impact of the Scots versus the Celts, he would have needed to multiply the amount of resources the Saxons devoted to each child and then used that calculation to estimate the child's survival chances as compared to that of an Irish child whose parents (peasants laboring in the service of English masters holding title to Irish land) were lucky if they were able to give their kids a potato, a bowl of porridge, and a bit of milk. Our ability with symbols has enabled us to create laws and governments that give us the power to declare that we own or have the right to control resources far in excess of anything needed for our immediate, personal survival. There *is* something unusual in human reproductive behavior but, as I will explain more fully in the final chapters, the solution to the "problem" Darwin describes lies not in regulating birth rates up or down depending on class, race, ethnicity, or geography but in correcting the flaws distorting the distributive function of our economy.

When designing my doctoral dissertation, I didn't have the benefit of the twenty years of economic activism that I now have and assumed, like Darwin, that the most successful men would have the most offspring. Unlike Darwin, I didn't impose objective measures of wealth onto the men my subjects described but accepted the individual woman's definition of success. However, some of the traits that triggered desire were correlated to material success, and, like Darwin, I assumed that women would have more children with men who were described in that fashion. I also assumed that women would expose themselves to pregnancy by being careless in their use of birth control with men who were, according to the woman's own assessment, successful. I was wrong in both of these assumptions and left graduate school, Ph.D. in hand, convinced that neither Darwin nor I knew nearly enough about human economic behavior. If the defining characteristics between chimps and humans could be traced to a mutation enabling concealed ovulation, where might I find the answer to the disparity in economic behavior between humans and all the other species on this planet? How was it that we were plagued by staggeringly high levels of inequality and red-winged blackbirds weren't? I spent the next two decades reading textbooks devoted to the dismal science and volunteering with foundations and groups concerned with addressing the disproportionate access to wealth that characterizes human societies. Getting this right was, and remains, about so much more than simply having fun debating the pros and cons of various evolutionary theories. Not only are we plagued by dangerous and growing levels of social and economic inequality, but our impact on the commons in nature is threatening many of the species on the planet, including our own, with extinction. It's imperative that we find a solution to this existential threat, and in order to do so, we have to take a clear-eyed look at the species causing so much of the damage. Understanding the evolutionary path that brought us to this place is not enough. We're left with the hard question: can we change?

It turns out that Darwin made two evolutionary miscalculations. One of these, which I have addressed in my introductory chapters, concerns speed. Darwin's early research interest was geology and his first publications were in that field. Through geological observations, he became convinced of two things: the deep nature of time and the gradual progression of changes in the earth's physical attributes over billions of years. Darwin, the geologist, took his understanding of the billions of years since the physical origin of our spinning planet and projected it onto the life forms that eventually developed. He was, of course, aware of the sorts of occasional, more abrupt, geological changes wrought by things such as earthquakes and volcanic eruptions and saw in them an opportunity for natural selection to seize on slight variations between one member of a biological species and another and to favor those traits that provided differential advantages, a process that resulted in an old-model animal or plant evolving into one that would be a better fit with its newly altered environment. Of course, his assumption was that fully incorporating these biological changes would happen at a slow rate of speed and might require the millions of years provided within the more typical and less-dramatic-than-a-tsunami rate of *geological* time. He failed to carefully consider the difference between the *rate* of change available to living creatures versus that determining the earth's physical structure. Geological change is imposed by outside forces acting upon inanimate structures. The flora and fauna undergoing evolutionary change have a bit more say in the matter than does a sandbar suddenly shifting position in response to rising sea levels. Plants and animals are alive. They have agency. Biological change happens at a different rate of speed than geological change.

Despite the fact that Darwin never argued that it would take species as long to form as it took the earth, his geological bias nonetheless contributed to his miscalculation of the potential rate of evolutionary speed. Biological change can happen fast, and it is not necessarily *ongoing*. Once an animal is well adapted to its particular ecological niche, its morphology and behavior tend to remain relatively stable. It was the failure of the fossil record to demonstrate the gradual change envisioned by Darwin that inspired Stephen Jay Gould and paleontologist Niles Eldredge to develop their theory of punctuated equilibria. Eldredge and Gould argued that species could remain stable for five to ten million years. However, when conditions changed, biological entities were able to respond fairly quickly. Eldredge's book, *Time Frames*, was published in 1985, and in it he suggested that a new species could evolve in five to fifty thousand years. As the examples cited in the early chapters of the book you are now reading make clear, we know that significant evolutionary change can happen in just a few generations. It goes without saying that Darwin was a brilliant scientist. We needed his knowledge of deep geological time in order to counter the prevailing religious doctrine that was impeding the advancement of science. It's completely understandable that, given the relative stability of species, Darwin would assume that evolved change happened slowly. He didn't have the advantage of being able to track DNA sequences in the lab, of being able to document changes in gene frequencies within just a few generations. In addition, the ecological interdependence of *all* species was not as obvious then as it is to modern researchers. Of course, Darwin saw the link between parasitic species and their hosts. Symbiotic relationships are likewise easy to document if one is using a fairly narrow lens. However, the notion that extinction at the bottom of the food chain can quickly ratchet its way up to the top is fairly recent. Remove one thread and the whole web of life is at risk. Darwin was concerned with adaptations and evolution in response to the various niches provided by the physical world, but animals don't simply adapt to their geographic space, they also *change* it and evolve in reaction to those changes. What was once a relatively inert planet has become one teeming with life. These life forms interact and adapt in a fairly harmonious balance. That's not to say species don't try to avoid being eaten by each other. In a way, the prey/predator relationship only serves to highlight the importance of their dependency. All individual plants and animals on this planet survive only through the death of something else. The soil that nourishes vegetative growth is composed of matter provided by the decomposition of plants and animals that have been transformed by death into a life-giving force. Nothing is wasted. Life feeds on death and begets life. But the degree to which humans have altered the planet is unprecedented. Our runaway consumption threatens to upset a balance that has been achieved over billions of years. The glacial pace of geological evolution no longer holds when human activities are melting the glaciers faster than any of us can adapt. The very nature of the physical planet and the life it supports are all shifting so rapidly that it's hard to keep up. We haven't seen this coming, at least in part because Darwin was convinced that in order to be evolutionarily successful, individual actors had to have lots of offspring. He didn't worry about the environmental changes being wrought by *Homo sapiens* as much as he worried that some segments of our population might be having too many children. However, especially in the case of humans, it's not the number of descendants we have that's important; it's what those descendants are doing.

It was getting led astray by Malthus that contributed to Darwin's obsession with numbers, another of his evolutionary miscalculations. A common measure of reproductive success begins with counting and ends with awarding the prize to the individual with the most offspring. But what works for other species doesn't work for us. The trick is in understanding that the true measure of success is not to be found in tallying up how many individuals get born but in how many of them grow up to become reproducing adults. Reproductive success is directly related to the ability of the parent generation to control enough resources to enable their offspring to survive to adulthood. In all species except our own, that economic ability is dependent on the size and strength of the parents, traits that are roughly equivalent between one member of the species and another. Human economic behavior differs dramatically from the economic behavior of all other species, and in determining reproductive success in humans, we must take those differences into account. If Bill Gates could establish title to no more of the resources necessary to life than he could defend with his own body, does anyone really believe that half of the wealth in the world would be controlled by only one percent of the population? No, of course not. The evolved human skill with language and technology means that some humans are able to translate accidents of birth, talent, and luck into the symbolic ability to lay claim to money, power, and other material and social resources far in excess of anything that could be defended by the physical strength of the fortunate few at the top. So, if we want to calculate the difference in reproductive success between one human being and another, it is necessary to do more than simply count and compare the number of children that those individuals have; we have to also figure the availability of resources that can be devoted to the rearing of each child. Whose children, grandchildren, and descendants far into any foreseeable future have the best chance of surviving to reproductively successful adulthoods? Mine or Mark Zuckerberg's? Please, Mr. Darwin and Mr. Greg, may I have your focused attention? Rich people have almost always been able to achieve greater reproductive success by having just a few children than poor people can. There's truth in that old adage, "The rich get rich and the poor get children." Historically, especially in agrarian communities, children were a form of wealth, a way to enhance a family's control of resources. In addition, when an individual child's chance of survival is low, it can be strategically adaptive to have enough children that at least some of them can be expected to survive to adulthood and carry the genetic line forward.

All living things strive to enhance their control of resources. Some species achieve that goal by maximizing the number of individual members representing their twig on the tree of life. If nothing more than measuring numbers was required to calculate the relative success of one *species* versus another, the ocean krill with a population estimated at 500 trillion would put our "mere" 7.7 billion to shame. There are species that pursue a different tactic and maximize their control of resources by increasing the size of the individual members. Elephants are a case in point, as are whales. Whales, by sheer mass, are the largest animals to ever populate the planet, heavier than any dinosaur that ever lived. Of course, in order to out-maximize even an elephant, they had to leave their land base and return to the sea. I remain fascinated by Michael Ryan's introduction of the seductive allure of *more*. But when creatures other than humans play the *more* game, there are built-in physiological limits. Lacking the sturdy legs of an elephant, whales headed back into the ocean so that their buoyancy could support their weight in a way that their legs could not. However, biological constraints come into play even when one is earning a living in the deep sea. The jawbone of the truly enormous baleen whale has a feature that prevents its mouth from opening so wide that the volume of water containing the amount of krill necessary to keep it alive does not, itself, swamp and kill the whale. Only human resource-accruing ability fails to be held in check by particularities relating to size and shape. And why is that? Because we alone increase our access to resources not by what we do *physically* but by what we do *symbolically*. Physiological constraints won't stop or limit our ability to ravage the natural world. It's going to take symbolic restraints to stop the over-consumption that symbolic control of the environment has enabled. In other words, it's going to take carefully designed governmental policies enforced by carefully designed laws. And in order for that approach to work, we have to figure out how to come up with solutions that are compatible with our evolved species-specific behavior. It won't be easy. Our language skills may have given us a future tense that evolution lacks, but that same talent for talk has enabled a tremendous amount of self-deception and deceit. In addition, we're stuck with brains that were shaped by concerns with reproduction and survival in the immediate here and now. We're not so great at honest calculations of still-on-the-horizon risk. However, since it is our economic behavior that is getting us into so much trouble, let's begin the search for remedies by examining the link between sex and money. Are there built-in limits to the seductive charm of symbolic expressions of *more*?

Chapter 12: The Economics of Desire

Shortly after the news broke the story on Jeff Bezos's potential lawsuit against the *National Enquirer* for threatening to print intimate photos and love notes he'd sent his girlfriend, Bill Maher succinctly summed up the situation by reminding Bezos that there *was* something in his pants that made women want to have sex with him, but it wasn't his penis; it was his wallet. If the current obsession with sending dick pics reveals anything about the male psyche—and I truly hope it doesn't—I think Jeff should be more concerned about the exposure of what was in those notes than he is with the display of his private anatomical parts. Shouldn't the richest man in the world have the smarts to figure out that it is only his extreme wealth that elevates such behavior above something that could land him in jail if he performed it in a public park? Don't get me wrong. I'm not a member of the anti-sex league, but I do worry about the pornification of flirtation made possible by the digital age. I'm just female enough to value sexual intimacy and to feel sad that today's young people of all genders may be missing a lot of the tender fun and sense of shared secrets and discovery that accompanies gradual revelations of personal truths and body parts. Having a selfie of somebody's penis show up on my phone would communicate more to me about the sender's tone-deaf self-adoration than it would about his desirability as a partner. I think Bill Maher got it right. Bezos shouldn't kid himself about the trait that makes him attractive to women.

This story reminds me of an example with roots in my days as a graduate student. One of the men in my group of Ph.D. candidates was a non-traditional student who had been married and divorced twice and had returned to school after having started his own quite successful business. Howard was a wonderful guy, a witty, accomplished man known for his generosity and love of women. A year or two after he left school, and while I was still struggling with my dissertation, Howard and I accidentally bumped into each other. At that point, we were both between marriages and immediately began sharing tales of romantic misadventures. He expressed surprise at my story of love gone wrong. "Not you, Heather. You're the courtship expert."

"Oh, Howard," I told him, "when it comes to men, I have an Achilles heart."

He laughed. "Funny thing, when it comes to women, I have an Achilles wallet."

I wish women weren't drawn to men with money, or at least not to the extent that they are. Even friends and acquaintances who tell me that they don't like being in relationships with rich men admit that a man's wealth captures their attention in ways they would rather it didn't. We *want* to love the pauper as much as we love the prince, but not even our fairy tales manage to find happy endings in plots where the good man who won the fair maiden remains poor. No. After she has proclaimed her undying love despite his apparent poverty, he reveals that he has simply been a prince in disguise all along. Kissing a frog turns him into someone with a prominent place in the royal household. Beauty loves the beast and the spell is broken. Of course, Beauty was already aware that the beast enjoyed a life of plenty. Was it only that which enabled her to see past his rough exterior? No matter, when everything was said and done, he had it all: money and dashing good looks. Of course, the poor servant girls, the Cinderellas, are always beauties in their own right. The plain librarian whips off her glasses, her hair tumbles down from its harsh bun, and—suddenly revealed as someone drop-dead gorgeous—she wins the love of the man she has secretly adored. Who is almost certainly rich. The ugly duckling was a male who was transformed by beauty, but while that may work for birds, in the world of humans, men are judged by their ability to provide and women by their looks.

It isn't fair, but those tales tap into something that feels uncomfortably innate. One could argue that women at least love men for traits that aren't inborn but represent characteristics a man can choose to devote time and energy to developing. However, if we really want there to be a level-playing-field component to the size of a man's wallet, we ought to work harder than we do to ensure that equal opportunity is not just a nice idea but is also something supported by our laws. And how about the physical attractiveness required of leading ladies in fairy tales and in real life? Not all of that is inborn. Women devote an inordinate amount of time and money to improving on the looks with which they were born. It's not just men who benefit from opportunities resulting from the economic advantages enjoyed by their parents. The various surgical, stylistic, and cosmetic interventions made possible by access to material resources also determine the face and figure a young woman presents to the world. Notice that adjective "young." When it comes to women, there definitely is a premium placed on youth.

Digital dating platforms make it possible to gather a wealth of data on male versus female preferences. Of course, the things measured by nothing more than tallying up the number of right swipes versus left are necessarily superficial, but one recent study offers little comfort to any woman over thirty hoping to find a love interest on such a site. At what age did women reach their peak of desirability, at least as measured by screen swipes? Eighteen. At what age did men acquire equal status? Fifty. What accounts for the difference? Those who turn to evolutionary biology for explanations, and I admit to being one of them, should brace themselves for overly simplistic and disheartening answers. A survival advantage fell to the children of women who chose men for their ability to provide, and the focus on wealth became enshrined in our species by sexual selection. What else could possibly make a fifty-year-old man more sexually appealing than one in his physical prime? Only the fact that it's less about sexual prowess than economic prowess. Men can be assumed to have reached a peak in their earning ability by their early fifties. So. How are we to explain men being drawn to younger women? It is hypothesized that the traits that trigger a man's sexual longing reflect those associated with females in their reproductive prime. A woman can be assumed to have hit the peak of her physical attractiveness at age eighteen: youth and beauty. The two nouns are so often joined, it's as if the phrase has become a single word.

What are we to make of it all? My takeaway is that our choices of romantic and reproductive partners are influenced by both our culture and biology, and only rarely are the aspirations imposed by either fully realized. Most women end up having children, and although the percentage of men who become fathers is slightly smaller than that for women, so do most men. We adjust to the realities of life as lived rather than to life as imagined. The most popular *New York Times* article of 2016 was a piece by Alain de Botton that went viral. The title? "Why You'll Marry the Wrong Person." The response seems to indicate that while this human institution may fall short of romantic expectations, its very endurance indicates the strength of evolutionary dictates. "Marriage ends up as a hopeful, generous, infinitely kind gamble taken by two people who don't know yet who they are or who the other might be, binding themselves to a future they cannot conceive of and have carefully avoided investigating." Who can really argue with anything described as hopeful, generous, and infinitely kind? The avoidance-of-careful-investigation aspect of it all reflects the manner in which evolution has shaped our brains to blind us to the possible costs associated with reproductive behavior.

Happiness is based on so very much more than simplistic stereotypes about wealth and beauty. In all things, I think we should resist evolutionary programming that conflicts with our personal goals, the ones we choose for ourselves. But I'm left with the troubling fact that control of material resources headed the list of traits that made a man reproductively attractive to the women I interviewed. Much as that bothers me, I'm bothered more when we pretend that something other than the size of a man's bank account is what made the heroine of the latest popular book or movie fall into the sort of head-over-heels love that fuels the dreams of those looking for happily ever after. Back in 2011, when *Fifty Shades of Grey* was first published online and then as a print-on-demand paperback, I had no desire to read the book, despite the fact that the aspiring author in me was impressed by the eventual book deal and resulting sales figures. It all sounded far too predictable. Another fabulously wealthy man wins the heart of a pretty, but poor, young maid. It was only when the third book in the trilogy was released, and I began hearing explanations for the success of the series that downplayed the wealth aspect and focused instead on the longing by women in an egalitarian 21st century world for a take-charge kind of guy that I picked up a copy of the first book. I had to find out for myself if the reason women were swooning was that they secretly missed the bad old days when all they really needed was a man to take care of them. Suddenly people were telling me that women wanted to be dominated, and I became curious. A few of the women I'd interviewed had admitted to being charmed by men who, without their asking, ordered for them in restaurants. Because I'd been surveying women about the traits of their male partners and hadn't specifically inquired about the exact nature of the sex between them, no one volunteered that she was into sadomasochistic relationships. But some of them probably were, at least as an occasional romp. This is, after all, a form of sex play that a number of people embrace. So I read the book and shortly after reading it, posted the following review on Amazon.

> Given all the hype devoted to the bestselling *Fifty Shades* trilogy, I am astonished that no essay I've read—traditional reviewers seem to sidestep actual reviews in favor of pieces best described as social commentary—has made what to me is an obvious point: this is not a book about sex; it's a book about money. And since economic security is a scarce resource for many women in today's job market, I'm not surprised by the popularity of the books.
>
> Author E. L. James is quoted in *TIME Magazine's* "Briefings" regarding the success of her multi-million-dollar publishing phenomenon. She claims that the enthusiasm with which women have embraced her books means she's "not a pervert." The underlying assumption is that the book is based on the author's own fantasies, erotic or otherwise. I maintain that her primary fantasy is an age-old one: a shy young virgin is rescued from a humdrum existence by a man with gobs of money. Christian Grey, E. L. James's fantasy hero, has in actual fact made James a millionaire many times over, all without her having to be humiliated by a single real-life episode of sadomasochistic torture. Her abuse at the hands of the literary elite is no doubt assuaged every time she deposits the latest royalty check into a bank account that surely has achieved multiple satisfactions far exceeding the author's wildest dreams. The bottom-line there is almost guaranteed to make James's inner goddess dance the merengue.

Talking heads of the female variety point out that today's woman comes home from a tough day in the workforce looking for the sort of dominance a take-charge kind of guy like Christian Grey provides. The supposition that she would endure sexual torture as part of the trade-off says more about current economic inequities than it says about the sexual desires of the average woman. Because let's face it, long before the reader is introduced to Grey's sexual fetishes, we are bombarded with evidence of his obscene wealth.

In the second paragraph of *Fifty Shades of Grey*, we learn that "the enigmatic CEO of Grey Enterprises Holdings, Inc." is "some mega-industrialist tycoon." After fifty-four pages of shades of economic excess—still no sex—Grey has a first edition collection of three volumes of *Tess of the d'Urbervilles* delivered to Anastasia. Just in case we fail to get the magnitude of this gift, our sexually innocent, but materially savvy, young protagonist lets us know that they are worth fourteen thousand dollars.

One can hardly call Anastasia easy. She and Christian don't have sex until we've all been subjected to one hundred and twelve pages of economic foreplay. We are smothered, not with kisses, but with detailed descriptions of the kinds of obscene trappings of wealth that—prior to the 2012 presidential election in which we learned of mansions containing elevators designed just to transport luxury automobiles—might have been as unimaginable to most readers as are the descriptions of the sadomasochistic devices—equally high-end—in what Christian Grey refers to as his playroom but that Anastasia more accurately calls the red room of pain.

I confess to having read only the first book. I find all the emphasis on money just too creepy. Christian marvels at Anastasia's lack of a gag reflex. I, unfortunately, have one. The growing divide between rich and poor in this country and worldwide turns my stomach. I hate to see it glorified in the books on my nightstand. There is nothing admirable about wealth that enables exploitation—sexual or otherwise—and that is far too often based on a ruthless disregard for the humanity of those at the low end of the economic scale. But in a world largely controlled by the one percent and the power that money can buy, I'm not surprised that James's books have inspired the imaginations of women everywhere.

As an anthropologist, I've argued that the link between sex and a man's control of what I coyly refer to as material resources has no doubt been with us since the first Early Modern Human offered a bit of roasted bison in exchange for sexual access. But wealth was more equitably distributed then. As a species, we've become entirely too clever with symbols for our own good. Where we once could control only what we were able to defend with our own labor, the extremely wealthy can now harvest the efforts of struggling fellow humans whose suffering is located conveniently out of sight in distant

> continents or in not so terribly distant neighborhoods. The fresh kill from a hunt could not be accumulated and stored in offshore accounts and Swiss banks. The value of the exploited, monetized and monopolized natural resources on which all life depends can.
>
> Women have always been sexually turned on by a partner who demonstrates genetic prowess and an ability and willingness to support the children who result from the union. Over forty million books sold and counting, the fictitious Christian Grey's wealth has stimulated such widespread erotic arousal that there is speculation that the *Fifty Shades of Grey* series will inspire a baby boom. This filthy rich sadist may have triggered the sex responsible for all those predicted babies, but I seriously doubt that he's going to stick around to write the college tuition checks. I think the publisher should stamp "Buyer Beware" labels on the dust jacket. And just in case anyone continues to doubt that *Fifty Shades of Grey* is less about the erotic turn on of bottom-spanking and more about the economic turn on of bottom-lines, all that person has to do is imagine an identical plot, same names, same characters, same behaviors with only one exception: instead of being a multi-billionaire, this Christian Grey is penniless and stays that way until the notably unfulfilling climax of the book. Suddenly, this is no longer an erotic romance—sadomasochistic or otherwise. Instead, it has become a chillingly creepy horror story with Christian Grey cast as no woman's dashing suitor but as everywoman's truly frightening stalker.

As a species, we are so good at language-enabled self-deception that we hide our obsession with *more* behind euphemisms. But if we are going to be successful at saving the planet, it's time to take the blinders from our eyes and call the behavior what it is: greed. And greed is not a trait that individuals of any gender would rank high on the list of traits to value in a hypothetical partner. In fact, I doubt that it would appear *anywhere* on *anyone's* list. The traits that early females found attractive in their reproductive partners may have helped make this affection for the trappings of wealth part of our species specific traits, but it's not only females who can be described as materially driven. Men are equally caught up in achieving the kinds of status that money can buy. Love of money is not a sex-linked trait. Darwin turned to the equal transmission of characters in order to explain how it was that women had managed to achieve a level of intelligence that he may not have thought was the equal of men's, but that was at least not as marked as the difference between the tails of a peahen and a peacock. Equal transmission of characters endows both men and women with the full range of species specific traits. Where male/female differences most dramatically assert themselves is in the behaviors humans employ to enhance their appeal to the opposite sex. In other words, people of all genders are given the same box of crayons with which to draw their experience. However, the resulting art reflects different cultural programming, styles, genetic variations between individuals, and the biological impact of the X versus Y chromosome. This is particularly true in the areas related to courtship and reproduction. I can't make the point often enough. Biology is not destiny; it's statistical probability. Assume that you are placing bets on the traits that first grabbed a given man or woman's attention upon meeting a prospective romantic partner. If you consistently bet that men would initially be drawn to looks and women to indicators of success, you'd sometimes lose a particular bet, but by the end of a day spent wagering, you would have made money.

Why does all this matter? Establishing the evolutionary origins of biological sex differences matters only if what we learn enables us to design solutions to current problems that are compatible with traits that sometimes can seem stubbornly innate. Our patterns of consumption fall into that category. As humans, we place a very high value on our symbolic and technological skills. It's a good thing that we're smart because we are going to need every bit of our brainpower and ingenuity to get ourselves out of the Anthropocene fix in which we've trapped ourselves. In order to do so, we may have to abandon our faith in the infallibility of our brains and accept the fact that we are not nearly as rational as we would like to believe. The scientific method was developed to *compensate* for the shortcomings of a brain that was largely shaped so that we might fool ourselves about some measurable "truth" concerning our prospective sexual partners and get busy with the business of having babies. It may be time to ignore our lying, reassuring brains and start believing the scientists whose careful measurements and computer-assisted projections cannot be ignored, not if we want our grandchildren—or even our children—to have lives that bear any resemblance to the ones you and I enjoy.

Our skill with the future tense doesn't extend very far into the *actual* future. The human capacity for accessing how things that have not yet happened are going to impact us is tied to language abilities that *co-evolved* with the brain. A system of female choice enabled early women to anticipate which traits in their male partners would help their as-yet-unborn children survive. Yes, female choice, the engine driving sexual selection, was concerned with future action, but it was for events that would happen in a fairly *immediate* future, primarily in the next generation. Women, and females of all species, are really good at marshaling a fierce reaction to any perceived threat to their offspring. Never get between a mama bear and her cubs. Human parents impulsively throw their own bodies between their kids and an oncoming car. Although language gives humans an ability to project many generations into the future, the emotional component of that save-the-children reaction quickly diminishes the further down the line we go. We would heedlessly, happily, abandon everything we own in the here and now in order to save our here and now children. However, our concern about future generations becomes increasingly intellectual and less emotional the more "greats" we affix to the word grandchildren. In its own odd way, it's a demonstration of the predictive power of kin selection as expressed in degrees of relatedness. In a demonstration of the mathematical calculation behind such tabulations of inclusive fitness, British geneticist J. B. S. Haldane joked that he would willingly die for two brothers or eight cousins. Cue Richard Dawkins and the selfish gene. A mother has ½ her genes in common with a biological child, ¼ in common with a grandchild, ⅛ with a great-grandchild. Since we've already discussed the failure of the numbers game to adequately describe human reproductive success, we can simplify things by asking how much investment a mother is willing to divert from an existing child in order to provide for a hypothetical one a few generations down the line. I'm afraid the notion of "a bird in the hand is worth two in the bush" may help us understand why it's so very difficult for parents to dramatically change their strategy of investing heavily in the children they now have in the hope of a better life for those a half-century removed from them.

Human patterns of consumption are rapidly rendering the world uninhabitable. The signs are already there—fires, floods, tornadoes, shifting patterns of heat and cold. We know that we consume too much. Plastic litters the oceans and plastic particles make their way through the food chain and into our bodies. How can we be so gifted at pretending it's not actually happening? Well, we do love our material possessions, and we are really good at self-deception. In the beginning, our ancestral Eves were selected to manipulate the environment that enhanced the survival of their children. Once that fondness for acquisition was enshrined in our biological response system, it got out of hand. Without biological limits placed on our symbolic control of resources, we seem to have lost the ability to practice restraint when it comes to behaviors that were initially adaptive. We humans ravage the natural world in an ongoing quest for the raw materials needed to manufacture the things we feel compelled to own. Are we running out of the rare metals used in increasingly sophisticated technologies? No worries, we've already set our sights on mining space. Why do we behave like this? Because procuring and owning stuff triggers the feel-good response that accompanies the satisfaction of our species-specific drives. We were designed to consume. Women are drawn to men who display all the trappings of wealth because, when the species was young, they wanted to ensure that their children would have enough to eat. However, when the ability to provide morphs into greed, into an environmental carelessness that threatens the survival of those same children, it may be time to use the intelligence that also defines us to get busy solving our rather immediate problems.

It's only when we begin to absorb the fact that climate change is real and that it's happening in the here and now in ways that may significantly alter the lives of children we already know and love that we finally get the attention of the general population. I don't think this awareness has arrived too late, but I am concerned that its arrival hasn't been accompanied by a greater sense of urgency. Unfortunately, the statistics on the damage already done and on the magnitude of the climate disasters facing us in the next few decades are sufficiently horrifying as to cause us to roll over in a collective belly-up of denial. An opossum pretends to be dead so that the predator threatening it will assume it actually is dead and leave it alone. However, unlike the opossum reacting to an external menace, when it comes to climate change, we are both predator and prey. The environmental damage we have created is the very thing that now threatens our survival. The monster we're trying so hard to ignore into non-existence is one of our own making. When it all feels too overwhelming to be true and the alternative to facing the facts can be turning a blind eye to the reality of what is happening to the planet, it is important that the Cassandras warning of the rapidly approaching danger do more than simply tell us to be afraid. If we are to be prodded into action, someone has to give us confidence that the threat can be reversed.

Happily, there are solutions. We *Homo sapiens* are economic animals and respond well to cost/benefit signals. We *Homo sapiens* are engineering specialists and are good at figuring out technological fixes. Our many skills enable the creation of algorithms that make it possible for us to calculate the true environmental costs of all the products and activities that we consider essential to our lives. If we can label food products with every ingredient and calorie, surely we can quantify the carbon cost involved in the manufacture and delivery of the sweater that has caught our eye or was needed to grow that avocado and transport it from the tree to our table. By combining our economic, intellectual, and technological talents, we can devise and develop environmentally sustainable economies. If the expense of repairing the climate damage associated with something such as air travel were factored into the ticket price, market signals would force the development of efficient and nonpolluting modes of transportation and create jobs in the process of doing so. If we had to actually *pay* for the damage we caused, would we be so cavalier about our consumption? Of course not, but all of this assumes that what we call free-market capitalism actually functions as such. What if it doesn't? What if those who are running this show have hijacked free-market capitalism and gradually replaced it with a form of corporate capitalism that has figured out how to privatize the profits and socialize the costs? A more honest description of our existing market structure would be corporate socialism. In the early days of the socialist movement, a popular slogan encapsulated the vision of a just society, "From each according to his ability, to each according to his need." Today's corporate capitalism or corporate socialism, call it what you will, has twisted the noble goal into something that could more accurately be summarized as, "From each according to his ability, to each according to his power."

Since all humans are skilled at self-deception, it's possible that companies whose lobbyists have directed policies that shift the burden onto the citizens of the world aren't even consciously aware of their failure to take responsibility for the environmental damage they cause. As long as a company has a healthy bottom line and the stockholders are happy, its executives don't face much scrutiny. There is a certain necessary blindness involved in doing business as usual, but at some point the magnitude of the climate catastrophe will make it impossible to ignore. We've been conned into complacency, but the time has come to face reality. The game has been rigged in favor of those who already have. They control an ever-growing share of the world's resources and along with it, have accumulated an ever-growing share of the world's economic clout and political power. They write rules that are supposed to provide a check on corporate behavior but instead sponsor legislation that enables much of the biodegradation threatening us. That said, many of us admire and aspire to be like the high-profile folks who've started companies that have landed them at the top of lists of the wealthiest individuals. The movers and shakers think of themselves as truly good people, and when we consider their philanthropic activities, the rest of us would have to agree. However, the human brain is better at fooling itself and us than we're able to acknowledge. It is not easy for anybody to admit that they may have enjoyed an unfair advantage. We all like to believe that we have truly *earned* the privilege we enjoy. But what if it's not true? What if our/their disproportionate share of the world's worth is due to a flaw in the *distributive* function of whatever economic system has allowed these inequities to grow and thrive?

Before we can answer that question, we need a deeper and more widespread understanding of how human economic systems really work. It's possible that by valuing growth over sustainability, we have created a kind of economic Ponzi scheme that can't go on forever. Perhaps saving the natural world begins with examining the role that *ownership* of the natural world has played in its destruction, particularly when the monopolists who hold paper title have figured out how to pass the costs associated with their privilege on to those who've been denied access to Earth's bounty. Allowing private ownership without an economic accounting of all the externalities associated with that license means that it's possible to make others pay the price for damage caused by careless use. This is no way to design an economy with a future. This is no way to save the planet. All Ponzi schemes eventually collapse. The earth is a finite entity, and it's being ravaged by a species with infinite desires. When we were a young species with a small population, our odd approach to ownership may actually have enhanced our chance of survival, but now it threatens it. Most of us are no longer hunters and gatherers. We're a mature species that has become entirely too ingenious at establishing property rights that give us a legal, but not necessarily moral, right to exclude others. Individuals can and do now claim far more of the commons in nature than any person can really need, and the symbolic and conceptual prowess that has allowed this to happen must immediately be put to a different use. It's time to harness the many facets of intelligence ricocheting through those big brains of ours in order to design systems that will, at last, put a stop to the insanity of insatiable desire.

Chapter 13: Talking Sense to a Species with All the Answers

I'm an optimist at heart. The ongoing joke in my family is that my glass is not only half-full, it is actually spilling over with the most delicious liquid I've ever had in my mouth. My daughter teasingly refers to me as the-girl-who-cried-*wow*. How is it, then, that when I sit down to read what I've written so far, it all sounds awfully grim? Why have I focused so much on women selecting for men who are good at controlling resources and so little on love, the central thing all people want from an intimate relationship? Because we aren't put at risk by a surfeit of love. It's not a reckless abundance of social bonds that threatens us. It is what we are doing to the planet that makes me wonder if our glorious run as a species is about to come to an inglorious end. We are too good for our own good at changing the physical world around us. And, *in the beginning*, we had to be. We were born of a rather significant mutation that came with some downsides. There were deleterious side effects that rendered us less biologically fit than the ancestral apes with 24 chromosome pairs whose genes were rearranged in the process of our creation. Natural selection could have stepped in and stamped us out, but the fusion that made us less fit in some ways gave us brains and ovaries that made us more fit in others. We pretty quickly figured out how to remove the fangs from natural selection. It was no longer red in tooth and claw. We cleverly had robbed it of its power of elimination. We'd figured out how to protect ourselves from the harsh challenges of the natural world. We did to the natural world what the natural world had once done to us. We changed it. We conquered it. Or at least we thought we had. But we may have been living on borrowed time, running up a debt that has now come due. Our activities have altered the world so much and so rapidly that it is no longer a very hospitable place for us or most other living things. Can we modify the behavior that has brought us to this place? Can we do it quickly and in a fashion that is compatible with who we are as a species?

Which raises the question: just who *are* we? For starters, we're pretty smart. Or at least that's how we describe ourselves. One thing for certain is that our brains are dramatically different from those of other species. It's not as much a matter of tallying up variations in size and shape when comparing ourselves to our closest primate cousins as it is about trying to figure out the *functional* differences that set us apart. Our brains have undergone neurological modifications that enable them to encode information in a unique fashion. Something in the way our brain is wired and organized has produced an animal that can talk. Other species excel at communication. We are the language specialists. We can and do chatter away about all kinds of things, past, present, and future. We not only share ideas verbally with those in our immediate vicinity, but we also have used our unparalleled symbolic skills to communicate via the written word. We exchange ideas, passions, and know-how with others of our kind around the world. We can and do tell others and ourselves the truth, and we can and do tell others and ourselves lies. This last little habit of ours can be seen as a downside that accompanies language ability unless we view it through the very narrow lens of facilitating reproduction. Even then, it lands in the asset column of the ledger only when we're evaluating outcomes that prioritize the evolutionary process over individual expectations. Since our language and brains co-evolved, the two are so intertwined as to be inseparable. What is an upside for neural capacity is also an upside for verbal capacity. If we're good at one thing, we're going to be good at the other. Since verbal aptitude allows for modification and embellishment in the process of its own enactment, it doesn't take long for those who practice it to become adroit at self-deception and deceit. These last two skill sets can make it hard to talk sense to this smarty-pants species of ours.

Although our species-specific traits—language ability, technological prowess, and economic complexity—are made possible by our biologically based neural and reproductive innovations, they become uniquely human only through the diversity of their manifestations. Their expression varies widely from culture to culture, but the drive to perform them is biologically hardwired. We are born wanting to learn to talk, but the language we acquire depends on where we spend our formative years. How about technological prowess? Some of us build dugout canoes, others design computer chips. Think of economic behavior as nothing more than a description of how we determine access to the resources necessary to all life. The economic complexity that makes it human is culturally inspired. Small hunter-gatherer societies were composed of individuals connected with each other through kinship ties and cooperative activities that encourage the communal sharing of resources. The subsequent invention of agriculture made large, permanent settlements possible, resulting in pressure to develop economic systems with a heavy emphasis on written laws establishing exclusive property rights. Once *symbolic* control of resources was formalized, dominance hierarchies based on concentrations of wealth and political power inevitably led to social and economic inequality. But no matter what form the cultural expression of our symbolic, technological, and economic skills takes, the successful enactment of these biological drives is rewarded and reinforced by a flood of endorphins.

The performance of biologically based traits is so emotionally satisfying that it can get out of hand. More, more, more. Blame Michael Ryan's book, *A Taste for the Beautiful*, for my obsession with the animal obsession with excess. Bigger, better, faster, richer. Language ability is the thing that makes the behavioral traits that set us apart from other species possible. We invent gods, religions, philosophical and moral codes, art, science, life-destroying weaponry, and life-saving medical devices. We take the things that other animals do to extremes. Anything they can do, we can do more excessively. We know no limits. We accept no limits. Do chimpanzees fight, sometimes to the death, over territory? Yes, but only humans have the capacity to level the world with the push of a single button. I've seen the photographs of Australian termite "castles" that are close replicas of elaborate human cathedrals. However, those termites are confined by an instinctual blueprint to but a single patterning, while human architects produce a seemingly endless range of styles and designs. The only constraints on the Frank Gehrys and Jean Nouvels of the world are imposed by the laws of physics, and they have mastered those so well that the resulting edifice can, to the uninitiated, appear to defy even gravity. Humans have a talent for regularly appearing to defy gravity. We send rockets blasting into space. How long will it take before there are as many satellites circling the Earth as there are stars visible to the naked eye from its surface? Where will it all end? Because unless and until we figure out how to impose some limits on ourselves, it is going to end somewhere. The space entrepreneurs of the 21st century are already contemplating how to mine meteors and distant planets for the rare metals that are becoming increasingly rare here at home. Does anyone really believe it's a realistic strategy to assume that when we have used up all of the resources we need here on Earth, we can simply move on to other worlds?

How have we gotten ourselves to the place where it seems reasonable to be discussing what we will do when we have exhausted the means of subsistence here on this planet? We've achieved this dubious goal by employing our species-specific traits with unmatched cleverness. In order to reverse the trends that have brought us to the brink of a climate apocalypse, we're going to have to use the very skills that got us into this mess in order to get out of it. Can we invent our way to cleaner air? Can we institute economic incentives that will reward sustainable solutions? Can we figure out how to stop kidding ourselves into believing that our self-interests are automatically aligned with solutions that serve the greater good? I think we can, but before we reach that place, we need to better understand the in-group/out-group dimension of our social behavior and find a way to make it work with, rather than against, our better angels. We'll successfully solve our problems only when we learn to make our *shared humanity* the source of the feel-good sense of belonging that is one of the rewards of group membership.

We've thrived thus far as a species because language has enabled us to draw on what is unique about each of us and make it available to those who have different skill sets. Our strength as a species lies in the fact that we are not all alike. We don't have to be. I don't have the technical skill and knowledge base of gene-splicing geneticists, but I am able to use what others have discovered to inform my own story-telling approach to understanding how the world works and how we function in it. My brain can't hold all the facts associated with engineering, art, music, geopolitics, chemistry, physics, etc. but by reading and listening to those with expertise in a particular field, I am able to find my way to authorities who help me knit their knowledge together with mine. Inventive geniuses build on the discoveries of all who have gone before them. I don't think a single animal could possibly contain all the problem-solving ability and adaptive capacity that humans have combined into an exponentially growing encyclopedia of communal expertise. What is it that really makes *Homo sapiens* so special? The potential located in our infinite variety. Try thinking of the human species as a single individual. The combined strength of that one animal would contain reserves of genetic material and learned specializations that no other species in the world could begin to match, not even if they were able to aggregate the range of adaptive talents of every single one of them into their own super-entity. There just isn't enough difference between the discrete actors to add value by combination. Humans are different. What we as a species have managed to pull off has been possible only because our accomplishments have drawn on a reservoir of choices unavailable to species with more limited genetic and cultural options. The eugenicists among us have missed the point. By wanting to breed, design, and engineer other humans in their own image, they are not practicing some God-like benevolence. Instead, they are *robbing* us of the variety that has made so much of what we admire about ourselves as a species possible.

The instinct to favor people who look and act in ways that are comfortably familiar almost certainly has its origin in the need to choose our reproductive partners from those belonging to the same species. However, those who continue to insist on a narrow definition of likeness remain lodged in the most primitive aspects of our ancient evolutionary history. I recently heard a discussion of genetic engineering that promised that humans of the future might be selected and "CRISPRed" into creatures with IQs above 1,000. My reaction? What a pity. What a loss. There's a real danger in selecting for a trait as narrowly defined as intelligence based on little more than an IQ score. Why not focus on generosity or resistance to disease? Who gets to make those decisions? Who gets to benefit from any supposed "improvement". It's pure folly to focus on a single, isolated trait. Have we learned nothing from those animal fanciers who selectively breed versions of a favorite dog with such exaggerated focus on an identifying feature, for example head size, that they end up with a breed that can "give birth" to puppies only via C-section? We don't know what physical and cultural capabilities humans may need in the future. It's vitally important to protect and preserve our infinite variety.

I don't think anyone of any political stripe will argue with me when I suggest that we have become as excessive in our partisan divides as we have in our rapacious exploitation of the environment: tribalism and greed. If we can figure out how to tweak the expression of these two human characteristics, I'll be more optimistic about the future. Solving our problems will necessitate coming up with policies that cut across local, state, and national borders and in order to do that, we need to learn to see ourselves as bound by our common humanity. Humans are not the only social species, but we are the only ones whose affiliates can consciously claim membership in social groups ranging from those with whom we have daily face-to-face or social media contact to those containing people we'll never know by name or meet and whose numbers must be tallied in the billions, 7.7 at last count. Although we are aware of the existence of fellow humans far removed from us in space and time, it's a bit harder to get us to care as much about those far-flung members of our species as we do about those who belong to the sub-groups that directly shape our sense of self. Despite potentially expansive definitions of humanity, we remain a fiercely in-group/out-group species. Like everything else we do, we seem to have taken determinations of group identity to extremes. We invent artificial distinctions to create an infinite number of categories.

We are an amazing and complicated animal, but we are still animals. Evolutionary reproductive dictates insisted that we mate with others of our kind. In order to do so, we had to be very good at recognizing and responding sexually to members of our own species. That's how we initially established the genetic boundaries that define us. There may be an evolutionary, species-recognition explanation behind our crazy need to create in-groups and out-groups, but like so much else that humans do, in the exercise of that particular drive, we have outdone ourselves and are at risk of doing ourselves in. Think back to the deep-sea diving Bajau people of Indonesia. They have a genetic variation associated with their skill, but no one would ever consider designating them a separate human species. We are all *Homo sapiens*. Because we humans are skilled at controlling and manipulating the environments in which we live, we've always been able to migrate around the climate zones of the globe, exchanging genetic material with so much enthusiasm that I find it impossible to view the notion of race as anything but another of those artificial human constructs.

Stephen Pinker says it best. In *The Language Instinct*, first published in 1994, he makes it clear that there is absolutely no scientific validity to the concept of race. "Eighty-five percent of human genetic variation consists of the differences between one person and another within the same ethnic group, tribe, or nation. Another eight percent is between ethnic groups, and a mere seven percent is between 'races.'" Why has it taken us almost three decades to grasp this most basic of concepts? Because we really do love the notion of tribes. We humans grant ourselves an easy path to unearned self-esteem by claiming membership in whatever group holds power. How about those who fall outside the artificial circle of superiority? An often *accurate* sense of victimhood cements their endorphin rush of tribal identity. Many of the markers we associate with our false categories of race are nothing more than weather-sculpted physical traits: skin dark with enough melanin to block harmful UV rays in sunny climates, eyelid folds to protect against cold, snow, sand and wind. "Race" may be only skin-deep, but it does allow for instant sorting into the comforting categories of us and them.

We got off to a roaring biological start by being able to distinguish in-group from out-group. Like any behavior reinforced by the evolutionary process, it is very emotionally satisfying to define ourselves as part of an in-group. What we once did genetically, we now do symbolically. We gravitate to those whom we perceive as being like us. We tell ourselves that belonging to our particular tribe—with membership variously defined in terms of religious belief, sports-based fandom, political affiliation, alma mater, brand loyalty, etc.—makes us superior in some central way to those who fall outside the qualifying parameters. I'm writing this in 2021, a time when the political climate is so toxic that our symbolic definitions of self have become sufficiently entrenched that they can override even the genetic similarity that was the original basis of this in-group/out-group nonsense. Family members have stopped speaking to each other. Thanksgiving dinners move forward only after they are designated as strictly politics-free zones.

I have always been rather casual in my description of *Homo sapiens* as an in-group/out-group species and viewed our tendency to sort ourselves in this way as "only human nature." I'm no longer as casual about it all as I once was. It is long past time for us to figure out how to rise above this particular bit of human nature. We have to recognize that just because we are convinced that our group is somehow more honest and better able to sort fact from fiction than those who oppose us, that belief is not necessarily an accurate reflection of some external truth. We need to be willing to consider that in order to comfortably establish ourselves in our corner of the ring, we can be pretty good at overlooking information and distorting even things that experts might document as having a pretty firm base in reality. How did we get to the place where otherwise intelligent people can talk about facts and alternate facts? We take shortcuts to sorting the good guys from the bad guys by looking at the political party of those who are describing the events at hand. Our trickster brains are so good at telling us tall tales about the evil motives of those in the other camp that they spare us the trouble of actually having to learn much about them. *Exclusion* becomes the lazy person's shortcut to the endorphin highs associated with group identity. Obviously, the vast majority of us are convinced that the beliefs we hold are the best ones. We don't deliberately choose to embrace ideas that we think are incorrect or immoral. However, I don't think I'm hopelessly naïve in my conviction that there is much more that unites us than divides us. The *extent* to which we are willing to assume the worst in others seems to me to be a relatively recent phenomenon, one which has been exacerbated by the development of digital technologies that enable us to rapidly sort ourselves into the comfort of identity silos. We get the serotonin rush, the cheap thrill associated with belonging to our particular in-group without having to do much more than simply dismiss those walled off in different bunkers as our natural adversaries. Who are those people we are so quick to label as "other?" Discounting those outside our circle doesn't feel like much of a loss when it takes little more than few clicks of the keyboard to find groups that will swell the numbers on "our" side. We are spared having to do the hard work of actually getting to know those in the opposite camp deeply enough to gain an empathetic understanding of how they came to the positions they hold. After all, how can there be an in-group without an out-group? Labeling someone else as an enemy can provide a superficial path to defining ourselves. The digital cloak of anonymity and the social-media normalization of hate have combined to create a poisonous brew.

Unfortunately, those digital platforms are about much more than simply enabling people to connect with like-minded others. The development of the technological hardware and software sites now in widespread use by the general public began with the discovery of how easy it was to get individuals to share personal information about themselves online. From there it was a short step to figuring out how to make money from this basic need for human connection. The whole enterprise was facilitated by copyright protections that were granted before we had a deep understanding of the complexity of this new technology and its potential impact on human life and social structures. Some of the early entrepreneurs now acknowledge that these programs were designed to be addictive in service of the profit motive. When we go online, we're targeted by advertising informed by the data collected about us as we share our lives with friends and family and search the web to find information and products that interest us. Making a purchase triggers the hunter-gatherer high that once accompanied a kill or the discovery of newly ripened fruit, but things have been turned upside down. Every click enables the tech giants to harvest *us* and sell our data. We've become the prey, the hickory nuts collected in a basket that will be made available to others for a fee. The trails left by our patterns of use and desire make it easy to track us down and manipulate us into making whim-based decisions to buy. At the very time when we ought to be reconsidering the wisdom of a Ponzi-scheme economy based on endless growth, consumption has never been more effortless-instant gratification. Like Pavlov's dogs, we salivate at the ringing of a bell, at the push of a button. It has all happened so fast. Digital innovations are especially proficient at exploiting innate behaviors that once enhanced an individual's chances of survival and reproductive success, but when carried to extremes, they can have very different consequences. The intent was not to sow hate and encourage us to consume our way into climate Armageddon. The intent was to build profitable companies that would make the founding innovators rich while also providing a rising tide that would lift all boats. How, then, have we come to a place where it is sea levels that are on their way up and the only boats that are being lifted are multi-million dollar yachts? Why is it that the fishing vessels of the working poor are getting swamped and those who operate them are unable to keep their heads above water? In large part, it is because our economic system has allowed monopoly ownership of the commons in nature. Other species don't have this problem. They have biological constraints that make it impossible for any single individual to control more resources than it needs in order to survive and reproduce. Lacking biological constraints, we humans must establish symbolic ones that will keep our greed in check. We have to figure out how to guarantee access to the material resources—clean air, pure water, and the land on which to earn a living, build shelter and procure nourishment—that sustain life for every species, including *Homo sapiens*.

In other words, we have to learn to share. It's easy for most of us to share with our closest relatives. If we're going to survive, we have to expand our understanding of the ways in which we belong, not just to our immediate family, not just to the human family, but to the interconnected web of life that inhabits the planet. Our survival depends on our learning to make nature's bounty available to everyone, to the whole, wonderful network that defines life as expressed in all its astonishing variety here on Earth. We must develop symbolic restraints that prevent the out-of-control greed of a single species, *Homo sapiens*, from destroying the ecosystems on which all life depends.

Chapter 14: Tangled Up in Economics and Evolution

The trouble with defining ourselves by our enemies is that it forces conflict and discord at the very time we most need to work together in order that life continues to flourish on this spinning orb we call home. I've heard it postulated that perhaps an invasion by extra-planetary aliens would force all the peoples of the world to unite in order to defeat a common enemy. I have trouble looking to Mars for my salvation via either the rare metals that space entrepreneurs plan to harvest there or the proverbial little green men in flying saucers. However, there is a threat to our continued existence that is real enough and immediate enough that it ought to rip the blinders from our collective eyes and get the people of the world to finally view ourselves for what we are, a single, deeply related and deeply connected species that will either work together to survive or not survive at all. The fact that this threat is of our own making does not render it any less real. We have created our own environmental monster, and we should be able to figure out how to bring this particular Frankenstein to heel. Why are we so unable to see the apocalypse heading our way? How can we break through our denial? First, we have to understand how and why we have evolved the way we did, and then we have to believe, really and truly believe, that we *can* change, that change is possible. It is. We are technologically clever enough to figure out how to extract and contain the carbon and other emissions that have been building up since the dawn of the Industrial Revolution and figure out how to generate energy from sustainable sources. Our technological skills helped get us into this mess; they can play a major role in getting us out. We're already making progress in that direction.

I wish we'd listened to Al Gore in 2006 when *An Inconvenient Truth* was first published. It's time to listen to him now and take him at his word. In a September 20, 2019 opinion piece in the *The New York Times,* the 42nd Vice President of the United States pulled no punches in his enumeration of the range, extent, and devastating consequences of the climate events that are currently destroying lives and economies around the globe. It's almost as if he was saying that all the things he'd warned us about had come to pass. The news in the early paragraphs was so dismal that I hoped readers of the *NYT* found the fortitude to push through until they reached the part containing Gore's more optimistic summary of the progress we've already made. Wanting to avoid the belly-up catatonia and sense of hopelessness that too often follow bleak environmental assessments, the summary below skips the grim initial paragraphs of Gore's essay and highlights the good news contained in the conclusion, the part promised in the final words of his title, "The Climate Crisis Is the Battle of Our Time, and We Can Win."

Gore is confident that we've already developed much of the technology needed to reverse climate change. Solar and wind are now "the cheapest sources of new electricity in two-thirds of the world." He predicts that in ten years these two energy assets will provide more cost-effective power than that available from existing fossil fuel plants. Some American companies are already announcing plans to close existing gas and coal plants and replace them with sustainable energy sources stored in ever-cheaper and more efficient batteries. By reminding us that the two fastest growing jobs in the United States are solar installer and wind turbine technician, Gore's essay anticipated President Joe Biden's April 28, 2021 message to Congress, in which he described his action on climate as a jobs creating program.

Hundreds of cities, states, and large corporations have a pledged goal of converting to 100 % renewable energy. Some have already met the target. The number of private and public and electric vehicles is increasing exponentially. Charging stations are seen as a necessary part of our infrastructure. Internal combustion engines are being phased out.

Gore credits farmers with leading an effort to focus on regenerative agriculture techniques that avoid plowing and work to build soil health and fertility through methods that sequester carbon dioxide in the ground. Using rotational grazing and diverse cover crops enriches the soil and protects against erosion. Massive tree-planting efforts are enjoying worldwide support.

Gore realistically reminds his readers that partisan politics stand between us and effective policies. Only by eliminating political infighting will we end taxpayer-funded subsidies such as those that encourage and reward our continued reliance on fossil fuels. He argues in favor of carbon pricing and economic incentives that reinforce a reduction in harmful emissions. His remarks remind me of my own concerns. Tribalism and greed. Is there no way around the most destructive of our species specific traits? Gore points out that "... 79 percent of American adults and 86 percent of teenagers believe, finally, that the climate crisis is caused by human activity," and puts his faith in their ability to use the democratic process to elect leaders who are actually dedicated to leading in the battle against what many see as the single most important existential crisis facing us.

We are making progress, no one can deny that. We have the technological skills needed to halt and even to reverse climate damage. The bigger question is whether we can master the behavioral skills needed to be clear-eyed in confronting the more problematic aspects of human nature. It's always easy to point out the ways in which "others" need to change what they're doing. Solutions will come only when we push past denial and realistically assess the environmental impact of our own "tribes". I trust no proposed solution that starts by insisting that somebody else has to change what they're doing. By continually describing the battle as one pitting "us" against "them", we simply reinforce the tribalism and greed that block the pathway to effective solutions. We are all in this together against something that won't be solved by resorting to political bias and negative campaign slogans. We need to believe that we share a common humanity. This is a worldwide problem that demands a worldwide solution. We have to join forces to do the almost impossible: rise above the flaws in our human nature and see ourselves not as completing factions, but as a global entity, a single species with a shared range of traits, some generous, others less admirable. This is a battle we absolutely must win. I think this is a battle we *can* win, but only if we are brutally honest in our assessment of the darker side of some of the traits that define us as human. Unless we truly believe we can rise above our more brutish instincts, we don't stand a chance. In truth, it's not too late, but we must act quickly. We need a balance between optimism and honesty. The danger in overselling all that can be done through technological innovation is that we make it too easy for folks to believe the problem can be solved without their having to do anything more than they've already done by forgoing single-use plastic water bottles.

It's naïve to think we can simply invent our way out of this. In order to win the battle, we're going to need to harness our political expertise as well. Money in politics is a major roadblock between us and effective solutions. As long as corporations contribute to campaigns, their lobbyists will write policies that shape laws in favor of those who want to continue to socialize the costs of their activities while privatizing the profits derived from careless environmental exploitation. Even individual contributions tend to encourage regulations that favor the wealthy. The amount of the political "gift" comes with tacit strings attached. It can be seen as a *quid pro* offer with an unstated but expected, *quo* reward. Not all contributions are self-serving, but the sheer size of some gifts can influence votes in ways that favor monopolists. This is especially true when the politicians in question may, themselves, have wealth they wish to shelter or enlarge through rent-seeking activities. There was a time when a political career was considered an opportunity to serve the greater good. I'm not in favor of forcing politicians to endure financial hardship, but neither do I want them to leave office enriched by extreme wealth of questionable origins. Legislative problem-solving would be more rational and efficient if all elections were publicly financed. In order to qualify, prospective candidates would be required to document the support of x-number of registered voters in the geographic area they hope to represent. Once in office, their salaries should be comparable to those in the private sector with similar responsibilities and talents.

Economic rent is income that is claimed through nothing more complicated than the ownership of a basic resource that others need. The entity holding paper title captures unearned fiscal rewards based on little other than the fact of the holding. The point is not to outlaw private ownership, but to put a stop to such rent-seeking behavior. Who would be willing to devote time, ingenuity, and money to utilizing a resource if their claim to the improvements erected on a piece of land, for example, were not secure? I don't want to deny anyone a legitimate profit when they have worked to productively employ the resources they hold. The trick is to separate out the gains that flow to title-holders as a result of their ingenuity and effort from the amount of return that is based solely on economic rent, the *unearned privilege* derived from the exclusion of others. In an ideal world, a world that American economist Henry George convinces me is possible, *all taxes on profit, income, sales, trade, etc., would be eliminated* in favor of one that would be levied on holdings in or damage to naturally occurring resources. George blamed growing income inequality on our government's failure to collect what he called a land tax—remember, he used the term "land" to refer to all the gifts of nature, not simply the ground under our feet— and what I prefer to call resource rents. Rent-seeking behavior is made possible by allowing these resource rents to be privately, rather than publicly, appropriated. George's theory is even more relevant in the age of climate change than it was when he proposed it in 1879. It's cause for celebration when an economic theory that was developed in pursuit of *economic equality* should also hold the answer to the climate damage that has resulted from unchecked human greed. There could be no more efficient solution to the twin problems of poverty and environmental destruction than collecting economic rent to fund government services and provide a Universal Basic Income to the citizens of the region governed.

If anyone continues to doubt the kinds of economic distortions and injustices that result from our failure to adequately price and collect, *for the public good,* the economic advantage that accrues to those with monopoly holdings, I refer them to an article in the October 26, 2019, *New York Times,* "Symbol of 80s Greed Stands to Profit from Trump Tax Breaks for Poor Areas." Michael Milken, otherwise known as "the junk bond king," is now out of jail and advocating for the establishment of opportunity zones. What is an opportunity zone? According to the official IRS website, "Qualified Opportunity Zones were created by the 2017 Tax Cuts and Jobs Act. These zones are designed to spur economic development and job creation in distressed communities throughout the country and U.S. possessions by providing tax benefits to investors who invest eligible capital into these communities. Taxpayers may defer tax on eligible capital gains by making an appropriate investment in a Qualified Opportunity Fund and meeting other requirements."

Despite the stated intent of encouraging economic development in distressed areas, in actual practice, these zones have primarily become an "opportunity" for the rich to become even richer, without having to expend much effort other than that involved in making contact with their well-connected cronies. According to the *NYT*, Mr. Milken "has investments in at least two major real estate projects inside federally designated opportunity zones in Nevada." Thanks to a bit of friendly lobbying, an earlier decision that considered these sites ineligible because the residents were too affluent was both reversed and expanded to include adjacent properties. So, legislation that was intended to help the poor has been co-opted in order to enable "legal" theft by the one percent. It's Robin Hood in reverse. The rich continue to get richer without having to lift a finger, unless one counts hitting speed-dial as lifting a finger. Thanks to the lobbying efforts of the industry, real estate "investments" enable financial killings over and above that made possible by monopoly alone. These windfall "profits" encourage a sense of entitlement, a belief that it's "smart" to have figured out how to get something for nothing. Engaging in such activity provides a training ground for future corruption. It's not only Mr. Milken's title to 700 acres of empty land just down the street from an intended enterprise zone that has given him so much economic clout. It turns out that he has also acquired approximately *700 other investment properties* in different locations around the United States. All this is occurring at a time when millennials—unable to afford to rent their own places, let alone come up with a down payment that would enable them to purchase—continue to live with their parents. Communities everywhere are struggling to address a lack of affordable housing. It's ironic that many of them look to enterprise zones as a way to remedy the situation. They'd be more successful if they simply removed all taxes on improvements and instead collected appropriately high property taxes on the locational value of the site. Rates levied on locational land values are not actually *taxes*. As a way of referencing their origins in the *unearned* income generated by *economic rent*, it would be more accurate to refer to these charges on land as *land resource rents*. Individual resource rents fall under the broad category of common-wealth rents, income that emanates from monopolies of all sorts but especially from those of natural resources: oil resource rents, water resource rents, mineral resource rents, etc. Collecting, for the public good, the full value of land resource rents would incentivize the creation of enough housing that people who work in a community could actually afford to live there. When passively withholding land from use becomes costly due to increased fees based on the *accurately assessed* value of the plot in question, a landowner will either develop the parcel himself or else sell it to someone with a more entrepreneurial spirit. The permanent removal of taxes on improvements would encourage far more development than any of the various taxpayer-funded schemes and payoffs that municipalities now use to entice new growth.

The ability of humans to *symbolically* control a disproportionate share of the commons in nature interferes with the right of every person to enjoy access to the material resources necessary to sustain life. This is the corruption of justice at the heart of both our careless use of the environment and our growing economic inequality. We must employ the very skill with symbols that brought us to the crisis point in order to make sure that no person or corporation is any longer allowed to get away with this kind of robbery. There is a way to guarantee that humans practice the same capacity to share and provide as that demonstrated by the various "races" of Juncos and other creatures who also inhabit planet Earth. The development of such policies is part and parcel of the effort to fairly compensate productive and innovative efforts in all segments of the workforce. In point of fact, the system I propose would no longer allow increases in productivity to be siphoned off by the rent-seekers among us and would guarantee that those actually doing the work would reap the rewards. We need productivity. We need innovation. What we don't need is behavior that enables some of us to get rich in our sleep while others have to work night and day simply to stay alive. It's easy to state the solution. Charge common-wealth rents any time anyone consumes a piece of the commons in nature. Whether the charges be on activities that pollute, consume, or monopolize, they all fall under the rubric of common-wealth rents. The resulting wealth flowing from them rightly belongs to us all and should be equitably distributed among us. We're rapidly running out of time and need to get busy figuring out how to put these policies in place. You pollute the air, you pay. You dump mercury into the water, you pay. You ravage virgin forests or overfish the rivers, streams, and oceans, you pay. You occupy a plot of land that someone else might also want to use, you pay. You control electromagnetic frequencies in ways that make them unavailable to others, you pay. You leave space junk orbiting Earth, you pay. And yes, those in the extractive industries would pass those charges on to the ultimate consumers of the product. And yes, that would make those products more expensive. And yes, that burden would be more acutely experienced by those at the lower end of the economic ladder. There is a way around this. These fees would be remitted to the government in the same way that we now remit existing taxes. In order to offset the unfair burden placed on people of limited means, those who already struggle to get by would qualify for a larger share of the remuneration through a process similar to the one that qualifies low-income workers to receive an earned income tax credit. The amount of this credit would be determined by a calculation of the common-wealth value of the resources made possible by nothing other than the bounty provided by nature or—in the case of land—by location and public investment and as the result of the development of neighboring properties. Landowners would be allowed to keep increases in value due to improvements that they themselves made, while any increases in value that resulted simply from a combination of location and the investments of others would be returned, through land resource rents, to the public that created them. The collection of common-wealth rents would also function as a way to recoup the rent-seeking advantage enjoyed by corporate monopolies of all sorts and would include charges reflecting the value created by public investment in the development and maintenance of infrastructure, including that associated with *digital* highways.

In a perfect world, this system would have been in place from the very beginning. In a perfect world, this return from common-wealth rents would be distributed equally to every single citizen. The state of Alaska already does this with their Alaska Permanent Fund. Every citizen, regardless of net worth, receives a yearly rebate based on the value of rent flowing from the oil extracted that year. Eventually, this is how the more comprehensive collection and distribution of resource rents would function. However, the existing inequities in wealth that have been created by allowing the private ownership of a public good are already so extreme that until those have been remediated, I find it appropriate to adjust the amount of the distribution in accordance with the need of the recipients. Once the system is fair and has corrected for the unearned advantage accruing to past and ongoing ownership of the commons in nature, then an equitable distribution similar to that practiced in Alaska is the ideal model.

Simply by being born, a person has qualified for the right to share in Earth's bounty. The total dollar amount of these rents is determined by calculating the value of all the natural resources and then collecting the portion of that value that results from monopolistic privilege. It may take a while to reach the point where we view all human beings as part of a truly global community. Until we get to that understanding, the value of common-wealth rents would be determined according to national boundaries and distributed among those living and working within the geographic region. The dollar amount of the economic rents flowing from natural resources is enormous, potentially enough to fully fund the government while also providing every human with a Universal Basic Income. *The goal is to eliminate all other taxes* in favor of collecting only the common-wealth rent associated with monopoly holdings of all sorts. Economic justice doesn't happen overnight, but by continually holding the vision of a fair system in mind, we will eventually achieve our goal. Having a conceptual understanding of the necessary components will keep us on track by providing a model against which to measure our progress.

Who can establish a fair claim to actually *owning* the land? Indigenous people occupied the territory that we now call the United States long before Europeans first set foot on the soil. Native Americans had no paper titles but practiced cultural patterns of use that provided predictable access. Their folklore offers a cautionary tale that serves to remind children from an early age of the dangers inherent in an inability to share. The Windigo of Ojibwe legend is a monster made fearsome by insatiable desires. It is a malevolent creature, characterized by traits—extreme greed, gluttony, and excess—that feel rather frighteningly descriptive of the current state of human affairs, one that tolerates huge disparities in wealth as "only natural." All we have to do is look at the adaptations that enable other species to survive without resorting to life or death struggles over food in order to realize that what are rare events among other members of the animal kingdom are entirely too commonplace among humans. When it comes to economic behavior, modern day *Homo sapiens* seem to be doing something incompatible with our own best interests.

How did my personal passion for Darwinian evolution lead to a journey that has left me tangled up in economics? Because Darwinian theories of evolution have *always* been tangled up with economics, or at least from the moment when Charles Darwin and Alfred Russel Wallace turned to the writings of the Reverend Thomas Robert Malthus for inspiration. Malthus was not a biologist. He was not even an amateur naturalist. He was an English cleric and early economist whose claim to fame was the demographic theory that human populations increase geometrically while the food supply increases only arithmetically. This assessment enabled him to see the world as a place defined by a struggle for existence in which the strong survived by outcompeting the weak. His theory was grounded in an observation of human behavior that was more *biased* than scientific in its data collection and that justified the exploitation of the weak by the strong. In fact, the good Reverend Malthus discouraged charity to the poor on the premise that their numbers would only increase if they had access to sufficient quantities of food. As we now know, that is not the primary factor influencing human population growth, at least not at the level of individual decision making. When humans are reasonably confident in their ability to provide for their families, birthrates don't increase; they decrease.

Malthus's distorted view found favor with both Darwin and Wallace because it provided them with a convenient, albeit *inaccurate*, mechanism that they could use to explain how species evolved. Sometimes our theories bring us closer to the truth, and sometimes they enable us to ignore it. A theory of evolution that continues to enjoy near-universal support and is believed sufficient to explain the evolution of all living things was based on the economic behavior of a single species, *Homo sapiens,* whose approach to securing access to material resources is completely atypical, an *anomaly.* Our symbolic control of and claim to the commons in nature bears little resemblance to the more equitable distribution practiced by other species. Mother Nature is bountiful. The scarcity that Malthus witnessed and that inspired his theory of population in humans—the only animal species he observed—is an *artificial* scarcity, one that is created entirely by human greed.

Chapter 15: It's a Bountiful World

My daughter Ingrid, my fifteen-year-old granddaughter Maeve, and I once took the Northeast Regional from Boston to New York to see my godson perform in the revival of *Kiss Me Kate*. From the minute we stepped on board, before the train had even started inching its way out of South Station, we were captured in our own little zone of happiness, bound together in a shared anticipation of good things to come. If only the United States had more rail lines and fewer highways. The train is such a civilized, stress-free way to travel. There's no worry about changing lanes, honking horns, angry commuters, and traffic jams. Perhaps coincidentally, it's also a fairly sustainable mode of transportation. The scenery on the route from Beantown to the Big Apple is mesmerizing, and the shores and marshes that morning in late February were filled with water birds and a spring-is-coming slant of light that warms tree trunks and sets the sap to rising. Mystic, New London, Old Saybrook, we skimmed past ancient seaport towns still bearing witness to their roots in the maritime trades. The three of us chatted and read and wandered between our seats and the snack car. On arrival in Manhattan, we checked into a hotel equidistant from the theater where James would be performing, Times Square, Fifth Avenue, and Central Park. The human experience, at least for those of us who enjoy a happy level of comfort and security, is cause for wonder. The New York skyline, whether viewed from an approaching train or looked upon from a city street, is dazzlingly incomprehensible. Who is the species that can do all these things, that can produce these architectural wonders? How is it possible that the same animal that gives us Frederick Law Olmstead and Central Park and skyscrapers and crisp hotel sheets and scented soaps and restaurants to please every ethnic palette and neon signs and shops filled with the answer to any conceivable desire has failed to provide for each and every one of its members? Much as the tourist bureau would like to keep the homeless conveniently out of sight, I can't help but wonder how it is that a species that seems able to solve any problem set before it has been unable to solve this one. But I push that troubling thought away and give in to celebrating these cherished moments with my family. That night in the theater, seated between my daughter and granddaughter and transported by James Lane and Cole Porter, I wept with joy, simply overcome by the human ability to create music, dance, art, and beauty. How is it possible to love this much?

Robin Wall Kimmerer is a botanist well-schooled in all aspects of scientific inquiry. She is also a member of the Citizen Potawatomi Nation and brings to her study of the natural world a profound sense of kinship with it. Kimmerer's scientific knowledge is enhanced by the richness of her indigenous cultural wisdom. She shares her two ways of knowing plants and animals through her teaching and in her writing as it appears in academic journals and books aimed at the general public. In Kimmerer's wonderful book, *Braiding Sweetgrass*, she quotes Marshall Sahlins.

> In an essay describing hunter-gatherer peoples with few possessions as the original affluent society, anthropologist Marshall Sahlins reminds us that, "modern capitalist societies, however richly endowed, dedicate themselves to the proposition of scarcity. Inadequacy of economic means is the first principle of the world's wealthiest people." The shortage is not due to how much material wealth there actually is, but to the way in which it is exchanged or circulated. The market system artificially creates scarcity by blocking the flow between the source and the consumer. Grain may rot in the warehouse while hundreds of people starve because they cannot pay for it. The result is famine for some and diseases of excess for others. The very earth that sustains us is being destroyed to fuel injustice. An economy that grants personhood to corporations but denies it to the more-than-human beings: this is a Windigo economy.

A few paragraphs later, Kimmerer continues her assessment and offers a bit of gentle hope by reminding us of our origins.

> Each of us comes from a people who were once indigenous. We can reclaim our membership in the cultures of gratitude that formed our old relationships with the living earth. Gratitude is a powerful antidote to the Windigo psychosis. A deep awareness of the gifts of the earth and each other is medicine. The practice of gratitude lets us hear the badgering of marketers as the stomach grumblings of a Windigo. It celebrates cultures of regenerative reciprocity, where wealth is understood to be having enough to share, and riches are counted in mutually beneficial relationships. Besides, it makes us happy.

Winter, spring, summer, fall. The seasons pass. I type these words on a sunny day in November. Earlier this week, Ingrid joined me in attendance at a Radcliffe lecture on tool use in New Caledonian Crows. When I was a little girl, one summer morning, a crow separated itself from a wild flock and practically fell from the sky and onto my father's forearm as he stood feasting on the uncultivated bounty contained in the heart of a tangled blackberry bramble. By the time Daddy got back to the house to wake his three sleeping daughters, he'd already given a name to the magical ebony bird that rode on his wrist with the practiced ease of a falcon. Joe clearly had been hand-raised by someone, but we never learned his history. It was enough that he showed up just after dawn every morning for the next few enchanted weeks and pecked on my bedroom window until my sisters and I came out to play with him. Sandy, Gail, and I quickly understood that a guaranteed way to get Joe's attention was to give him something with moving parts. He was an inventor, a thief, a playful scamp. We had a stone patio outside our living room door, and every morning we hauled things out there to see what antic behavior they might inspire. Joe watched Gail insert a pencil in a sharpener of the sort most often found securely fastened to a teacher's schoolroom desk. A quick student, he immediately seized the pencil with his beak, inserted it, and attempted to turn the handle. Joe never quite managed the circular motion that would produce a precision point, but he could effortlessly bounce a small, hard rubber ball across the patio. It is only with the hindsight lodged in my adult self, a person consumed by an interest in language and its relationship to the evolution of technological skills, that I give adequate due to what was probably Joe's most impressive feat. We gave him an empty Good & Plenty box, the kind with flaps that opened and closed at either end. Joe examined it, poked it, moved the parts that could be moved, then grabbed the small container in his beak and flew low and quickly around the house to a gravel driveway that was not visible from the patio. The three of us followed as Joe must have wanted us to do because he never soared out of sight, never rose above our heads. Once we caught up to him, he paused, checked to make sure one end of the box was securely closed and then dropped a half dozen or so pebbles into it. After he had fastened the top flap into place, he picked up his new toy/musical instrument and shook it. How did he manage to do that? How did he hold awareness of the gravel driveway in his mind when it was not possible to see its stones from the spot where we had first presented him with the box?

Were we never tempted to keep Joe, to contain him in one of the many large cages piled in an empty stall in the old barn at the end of a long lane that curved through the property? Of course we were tempted, of course we were. We discussed it. We wanted to make Joe ours forever. My sisters and I were animal lovers and had hand-raised countless orphaned critters, both wild and domestic. But confining Joe would have felt too much like a denial of the gift he'd made by choosing to wing his way into our lives. Although we were amused when he stole the shiny lens cap from our father's camera and stashed it among all the other bright objects he'd cached in a knothole high in the trunk of the black walnut tree that stood tall beside the house, we just could not bring ourselves to rob from Joe by thwarting his spontaneous desire to share his mornings with us. And, of course, there also was the matter of the implied judgment we would suffer in the eyes of the wild flock that circled overhead, cawing raucous warnings for the entire hour or so Joe spent with us each day. We didn't want to prove them right. They grew silent only when he tired of our games and lifted up to rejoin them.

Ingrid brought with her to the New Caledonian Crow lecture a seventy-year-old black and white photo of Joe, my sister Sandy, and me that Daddy must have taken on the first day of that miraculous bird's arrival. Joe fills the frame. I am crouched behind him, my hands tightly clenched together in an astonished prayer of disbelief and wonder. The sheer magic of it all! How is it possible to love this much? Why do I continue to love this much even though I learned early that love wounds in exact proportion to the depth and breadth with which it fills the heart? Because, here's the thing, the story of Joe the Crow has a sad ending.

Although my father had spent his childhood on a farm in Berks County, Pennsylvania, our little family of five always had—at least until we rented the ten-acre property where Joe found us—known a suburban existence. Daddy took the train to his job with an advertising agency in Center City, Philadelphia, while his daughters enthusiastically took to populating those neglected lawns and fields with all the creatures we thought necessary to a proper farm. The Gregors were our closest neighbors, and *they were* proper farmers. Although Mr. and Mrs. Gregor were older, closer to our grandfather in age than to our parents, they had a son, David, who was my classmate in the local public school. The first thing Mr. Gregor did when we moved in was offer to use his tractor to plow a spot where Daddy could plant some field corn to help feed our ever-growing menagerie of sheep, goats, a pony, a horse, a pig, chickens, and ducks. Our more urban assortment of dogs, cats, guinea pigs, rabbits, hamsters, and turtles had other dietary preferences. When it came time to harvest, Mr. Gregor showed up with a corn-picker attached to his tractor. He was a good, good neighbor. We tried to reciprocate by helping with the haying, but only Daddy and Sandy were strong enough to be much actual help. I desperately wanted to pattern my movements on David's relaxed competence, but failed on every count. Gail and I played at stacking bales in the haymow and came home covered with enough scratches that we were able to pretend that we had spent the day hoisting the prickly bundles into place.

That was a charmed and golden summer. And then one morning, Joe failed to appear. Assuming I'd slept so soundly that I hadn't heard his insistent peck, peck, pecking on the windowpane, I vowed to never let that happen again. The next night, I went to bed early, woke before dawn, and waited, and waited, and waited, but Joe had vanished from our lives. We never saw him again.

Heartbroken, I shouldered the blame, certain that my failure to respond had sent him off in search of more reliable friends. I chastised myself for not being more imaginative in my choice of objects with moving parts. Perhaps we'd bored him, and he'd traveled on in search of other kids. Maybe those children had put him in a cage. Maybe he was held captive somewhere, desperately trying to master the lock that stood between him and his freedom. A few weeks passed, and then one Saturday morning, Daddy walked home from a visit with Mr. Gregor looking somber. He gathered Sandy, Gail, and me around him. After a long pause, he sighed and then, apparently not knowing how to sugarcoat the facts, just stated them. "Girls, I have some bad news. Joe is dead."

We wailed in disbelief and, in between our keening cries, we badgered our father with questions. Early one morning, Mr. Gregor had headed for the henhouse to let his chickens out for the day. In this old farmer's telling, a rabid crow had swooped down at him. Sandy, in a desperate attempt to rewrite the inevitable ending to the story, cut Daddy off. "Crows don't get rabies!"

"Mr. Gregor doesn't know that," our father said quietly.

"Birds don't get rabies!" Sandy was fierce in her determination to stop this tale in its tracks. "Crows don't get rabies. Only mammals get rabies!"

Daddy gently continued, "Mr. Gregor was simply trying to keep his chickens safe." He hesitated. "He went inside and got his shotgun."

"Noooooooo . . ." We howled in unison and then dissolved into our separate, more individual, approach to tears. "No, no, no," we softly whimpered. "No, no, no."

But it was true. We inquired about the possibility of a burial. Not all eye-dropper-fed animals survive, and the three of us had made an art of officiating at respectful funeral ceremonies. But too much time had passed and apparently, drifting black feathers had been all that Joe left behind. Unless one counts the memories. Those are still very much with me.

Those ten acres on North Wales Road boasted a series of reed-filled ponds. I had a secret spot in the woods downstream from our house where I sometimes went to nurse my sorrows. My face must have still been blotched and swollen a couple of hours later when I rejoined the family because Daddy pulled me aside. "H," he said, using my childhood nickname, "nothing lives forever. Everything dies eventually."

"Not Joe. Not my Joe. He was too young. He was too special."

"He was," my father agreed.

"How could Mr. Gregor not know that? Mr. Gregor knows everything. How could he not know about the rabies?" My accusatory question was a bit dramatic. I had myself only learned the bit about mammals and rabies a few hours prior. Sandy was three years older and had taken a biology class that gave her all kinds of dazzling information that she sometimes, but not always, chose to share. "I wish Sandy had told him. I wish we could have figured out a way to tell Mr. Gregor the one thing that he really needed to know, the one thing we really needed him to know."

"No one could have seen this coming. Mr. Gregor didn't mean to make you sad. You know that he is a good man, a kind man, a good neighbor. You know that, right?"

"I know. I know he is. I just wish we could have figured out a way. I just wish I had told David about Joe."

I guess there will always be situations where we wish we had handled things differently. Humans have a future tense, but we don't always have a lot of future sense. However, there are some things we ought to be able to anticipate. Sometimes the warning signs are already in place. The trick is to determine what information we need to gather and what insights we need to share in order to circumvent a loss too great to be endured, a loss that would do more than simply break hearts, a loss that could actually stop them from beating. I want to be optimistic about the future of this miraculous planet of ours, but it's way past time to take the blinders from our eyes. We're in this thing together, and we're doing something wrong. No one person has all the skills we need to make the necessary changes. We have to work as a team. Each one of us must do our part. What do I envision my role to be?

If there was one thing, one single bit of information I'd want to share, what would it be? I'd want the world to know, to truly and deeply understand that the way humans control and consume natural resources has become maladaptive. Marshall Sahlins had it right. Robin Wall Kimmerer had it right. There is a Windigo raging unchecked among us. What we are doing is unnatural. We're creating artificial scarcity in a bountiful world. The solution does not lie in abandoning our marvelous technological skills. We need those skills and will reinforce the best of them by encouraging *genuine* productivity. Simply producing more, more, more and consuming more, more, more is not an accurate measure of economic health. True productivity occurs when we achieve the same results by using less of the natural resources that are a necessary ingredient in all manufacturing. Charging resource rents discourages waste and makes choosing renewable energy sources the only profitable outcome. The world is a bountiful place. It's up to us to figure out how to share it so that the bounty continues and is made available to all the species down through all the ages and all the generations to come, a world without end.

Chapter 16: In Summary: Forty Acres: An Evolutionary Insight into Social Problems

Can the nine and a half minute video of the callous murder of George Floyd really be all it takes to rouse white people from our slumber, from our blissful blindness to the lived reality of what 400 years of social and economic inequality has inflicted on those whose skin just happens to be a different color? Maybe white people need to demonstrate our "wokeness" by doing more than holding up signs declaring our solidarity with the Black Lives Matter movement. Maybe it's time to acknowledge that the main thing about white privilege is not only that it is unearned but also that it is stolen and has deep economic roots that have yielded compounded magnitudes of injustice, degrees of social and material inequity so vast as to be difficult to comprehend. Unless one has lived them.

White privilege attaches itself to us, not because of anything we have worked for or accomplished but simply because we happen to have been born with a tiny scattering of genes related to the degree to which our bodies produce melanin. That's it. The ancient atrocity, the rot at the core, is based on something that is literally skin deep, that has no biological reality and is unrelated to to any other trait—intelligence, strength, generosity, kindness, creativity, technological prowess, health, insight, compassion, etc.—that humans claim to value. How is it possible that the species that has dubbed itself *Homo sapiens*, the wise hominin, has gotten away with indulging in a cruel and flawed interpretation of evidence that is completely at odds with our more usual definitions of humanity?

None of us wants to believe that we haven't fairly earned the riches we enjoy. It is painful to acknowledge that we may not be as clever, hard-working, and deserving as we've convinced ourselves we are. A certain degree of self-confidence has survival value, and so we *Homo sapiens* have been shaped by evolution to be skilled at lying to ourselves. No person of any description finds it easy to admit that their good fortune is grounded in theft, but white privilege is. There is, of course, the obvious theft of personhood from those who were kidnapped, enslaved, and sold by profit-driven others who were able to justify the heinous nature of their actions only by pretending that those they abducted were not truly and fully human. This criminal bit of self-deception was made easier because there was a visible physical identifier, skin color, that enabled the enslavers to sort human from non-human in the literal blink of an eye.

Unfortunately, this behavior is not something that started in 1619. Members of our species have been enslaving each other almost from the moment when the adoption of an agrarian way of life permitted those at the top of the social and political hierarchy to declare themselves the "owners" of the natural resources—land, virgin timber, clean water, mineral deposits, unpolluted air—on which all life depends. From the original sin that sanctioned *symbolic* ownership of the Commons in Nature, it was but a small step to figure out a way to extort the labor of people who, "legally" denied access to nature's gifts, could then be forced to work the land without being allowed to retain for themselves anything beyond the minimum required to stay alive and able to work another day. Chattel slavery. Wage slavery. Both involve robbery, but only the former legalizes treating one's fellow humans as actual property.

As for land, who can be said to really and truly own the land? Trace any formal title back far enough, and you'll find that it originated in theft of one sort or another. However, there is something about land ownership in the United States that renders the centuries' old scars of slavery as perpetrated in this country particularly egregious and difficult to repair. If one wanted wealth without work, historically there were two obvious options: buy and sell land or buy and sell people. The U.S. was founded by individuals who were practiced at doing both.

After the colonizers had divided up all the land in their countries of origin and rendered it artificially scarce through legally defended claims, they sailed off to find new worlds to commandeer. What is now the United States was thinly populated by indigenous people who employed the egalitarian approach to land use common to hunter-gatherers. Given our propensity to deny the humanity of those we are about to kill or from whom we are about to steal, the Native Americans didn't stand a chance. Once the irksome "savages" were either murdered or pushed conveniently out of sight, we could kid ourselves into believing the newly discovered land was pretty much free for the taking.

King James I formed the Virginia Company of London in 1606, but forging a life in the new world was not for the faint of heart. In order to encourage productive development of the region's natural bounty, the Virginia Company offered any adult male with the means to establish himself in the unknown territory 50 acres of land. The wealthy industrialists who were able to take advantage of the offer had already mastered the trick of profiting from someone's else's labor. They banded together to secure large tracts of land, but in order to extract wealth from those plantations, they would need a large workforce. Since there simply weren't enough people for the kind of production they had in mind, those looking to enhance their fortunes without dirtying their own hands had a problem. Even if there had been a sufficient number of industrious, fresh-off-the-boat souls, what was to stop the new immigrants from simply heading into the wilderness and harvesting nature's bounty directly? Hunting, trapping, fishing, lumbering, mining—for those with a pioneering spirit, the means to the good life would have appeared to be suddenly at hand. When there aren't many people populating a territory with abundant "free" land, when the greedy can't force low wages by owning all the available land, an obvious—and odious—solution is to take to buying and selling their fellow humans. The first enslaved people began arriving in Virginia in 1619.

Injustice has been part and parcel of who we are from this country's not-always-entirely-honorable beginnings. Recent events have finally inspired what seems to be a near-universal agreement that it's time to bring the compounded inequities to an end. Let's start by comparing the amount of wealth held by white families to that held by Black. *The Boston Globe* Spotlight Team recently reported that in Boston, the median net worth for non-immigrant white households is $247,500. For U.S. Black households in Boston, it is $8. Net worth was determined by subtracting debts from assets. The most significant reason for the gap between the two populations is related to the difference in home ownership, a disparity that can be traced to active discrimination—past and present—against Blacks hoping to purchase homes or real estate of any sort. As Noah Smith points out in a May 22, 2019 Opinion piece in *Bloomberg*, "The ownership disparity is . . . responsible for a big share of the racial wealth gap."

Why is that? Houses don't increase in value, at least not without a lot of work. As any homeowner can tell you, a structure is no sooner erected than it starts depreciating. Without constant attention to upkeep such as painting, roof repair, window replacement, plumbing maintenance and general refurbishment, the buildings themselves lose value. In point of fact, the asset growth associated with home ownership has little to do with the home and is pretty much a function of the increase in the price of the land under it. I own a condominium just outside of Boston. In Massachusetts, condominiums are not assigned a land value for property tax purposes. Make no mistake, regardless of what it says on my tax bill, the dramatic increase in the estimated selling price of my condo in the decade or so that I've owned it has nothing at all to do with any improvements I might have made. I haven't so much as repainted the living room, but the value of my holding has skyrocketed. My asset has practically doubled while I slept, and that increase is almost entirely due to the locational value of the land on which the condominium stands.

Mark Twain is credited with saying, "Buy land, they're not making it anymore." Which brings me to those forty acres. General Sherman's Special

Field Order 15 proclaimed that enormous parcels of former Confederate lands should be broken into forty-acre tracts and distributed to the newly freed people. The enactment of this order almost immediately resulted in the settlement of 40,000 freedmen on 400,000 acres of land. Unfortunately, Andrew Johnson, Lincoln's successor and a southern sympathizer, overturned the Order in the fall of 1865 and returned the land to the planters who had once owned it. What if the Order had been permitted to stand and those who had been kidnapped, bought and sold, and brought to this country against their will had been allowed the same right to land ownership enjoyed by those with white skin? Do you think we would still see the disparity in wealth that is evident, not just in Boston, but everywhere we look in the United States? Not only did the freed people fail to retain those forty-acre plots, but they also were blocked from participation in the Homestead Acts, a federal government program that gave away 246 million acres, almost 10 percent of the total landmass of the nation to more than 1.5 million white families, native-born and foreign. These statistics and dozens of equally enraging examples are documented in *The New York Times Magazine's* August 14, 2019 special feature on slavery and leave little doubt that achieving economic and social justice is impossible without all people having the right to property in land.

It's not enough to right past wrongs; we have to figure out how to make sure they never happen again by addressing the root cause of growing inequality, not just in the United States but worldwide. Structural changes in the way we view and tax land holdings are long overdue. It's unrealistic to consider outlawing titles that establish the right to own land or any of the gifts of Nature. Nobody would plant a crop or build a factory without some guarantee against its seizure. If, as I argue, no one can be said to have a moral right to own the Commons in Nature, how can we earn the right to a title? By recognizing that with the granting of every fee simple deed comes an obligation to make payments into the public coffers, in perpetuity, for the right to that property's use. Economists acknowledge the existence of Economic Rent, the portion of any given holding's value that is unearned and accrues to the title bearer through nothing more complicated than his command of a basic resource that others need. The source of my suggested "in perpetuity payments" to the public would be drawn from the Economic Rent associated with "ownership" of the parcel in question.

It is inaccurate to call the fees originating in Economic Rent taxes; we'd be more on target to refer to them as Common-wealth Rents, wealth that rightfully belongs to all of us. It's especially important to use precise language when we're talking about the property tax, which is—with good reason—the most hated tax in the United States. The property tax is highly visible and the portion that falls on improvements actually *is* a tax, a burden and disincentive to the very kind of productive activity communities want to encourage. In most municipalities, the local property tax is the primary source of school funding. Disparities in educational opportunity perpetuate racial inequality. The more money a district spends on its schools, the better the schools, the better the schools, the higher the real estate values, and the higher the real estate values, the more likely we are to find functional racial and economic segregation. Public schools should be funded nationally, not locally. By having the federal government collect the sum of the Common-wealth rent associated with land ownership and distribute some piece of it nationwide according to student need and enrollment, we would begin to correct racial inequities in access to education. In the process, we'd also address the lack of affordable housing.

Residential homeowners who feel that I cut a bit too close to the bone by arguing that they should assume the burden for the nationwide cost of education have a point. However, they can relax. When it comes to funding public services, it's the owners of corporate, commercial, and investment properties who have been getting away with murder. For example, McDonald's has nearly $30 billion in real estate holdings. Given the power of the real estate lobby, the value of much commercial and investment land is under-assessed. Therefore, the pursuit of justice starts with accurate assessments. Since land value is largely locational and land can't be moved and hidden in offshore accounts, this is less onerous than assessing the amount of income derived from improvements. The popularity of LLCs can make it difficult to determine actual ownership. That's okay. The taxing authorities may not always know who owns these plots, but they do know where to send the bills. Given Mark Twain's reputed sound-bite summarizing why buying land is a strategic move, no one should be surprised to know that purchasing U.S.efdc real estate is attractive to foreign investors.

According to a 2019 interview on *NPR*, nearly 30 million acres of U.S. farmland are held by foreign investors. *Statista* reports that in 2019 alone, "property sales to foreign buyers totaled 78 billion U.S. dollars," a statistic that's hard to fully document since not all states require information on the citizenship of those who purchase real estate. As a 2017 article in *The Washington Post* pointed out,100 wealthy families now own as much land as all of New England, while "Eighty percent of us live on 3 percent of the United States." The estimated value of land in New York City alone is just under two trillion dollars, and—given the clout of the real estate lobby and legislation that allows for land to be depreciated along with buildings—that estimate is almost certainly on the low side. With statistics like these at hand, can anyone fail to grasp the importance of fully and fairly collecting the Annual Resource Rents associated with land holdings? Only by making the users of natural resources pay, will we finally begin to address a root cause of economic inequality.

In the United States alone, the collection of common-wealth rent from land would be more than sufficient to fund the education of all our citizens, from preschool right on up through college. There'd even be enough left to establish Universal Basic Income grants. Such guaranteed payments are often suggested as an important part of any move in the direction of reparations. The Alaska Permanent Fund provides a working example of how such a system could be implemented and financed. Every year, the State of Alaska grants each of its citizens a rebate based on the value of rent flowing from the oil extracted in a designated twelve-month period. The amount varies according to the price of oil and the whim of the State's elected representatives. In 2015, every man, woman, and child in Alaska received just over $2,000. The estimated worth of the grant in 2017 was $2,300, but legislators reduced it to $1,100. The payment in 2019 was $1,606. And that amount is only the income derived from the Economic Rent of *oil*. Just imagine the size of the potential pot if we annually collected the rents associated with all natural resources! The staggeringly large value of the total would be Common-wealth rent that rightfully belongs to all of us and should be fairly distributed and shared.

This imperfect union of ours was founded on more than theft; it was also founded on ideals of equality, liberty, and justice. These are noble—and achievable—sentiments. In the midst of troubling times, it can be hard to hold to the optimism expressed in the belief that the arc of history bends toward justice, but progress in that direction is reflected in shifts in the language we use. All *people* are created with equal rights to life and the bounty that nature provides. A democracy gives those it governs a powerful tool for righting past wrongs and ensuring that it lives up to its stated goals. We need to educate ourselves. We need to vote. We need to engage with our politicians and study their policy positions, and we must make certain that our elections are publicly funded so that no one in any branch of government is more beholden to their personal dreams of economic gain and the dictates of their donors than they are to the fundamental rights of those for whom they govern. We need to listen to each other and make sure all voices are heard.

Most of all, we need to acknowledge that as fellow human beings we share the curse/blessing of brains that enable us to turn a blind eye to the complicated tangle of factors motivating our own behavior and beliefs. There is a deeper truth than we may care to admit in the conclusion drawn by cartoonist Walt Kelly's Pogo, "We have met the enemy and he is us." But we are so much more than an acquisitive, in-group/out-group species. Humans are also capable of astonishing feats of generosity, kindness, and cooperation. We rightly celebrate our scientific and technological achievements, our works of art, literature, and music. Let's join forces to turn our backs on those aspects of human nature that have not served us well. Let's push tribalism and greed aside and nurture our better angels. We shall overcome, we can overcome, we *must* overcome, not just our "racial" divides, but the shadowy, self-serving underbelly of our evolutionary roots. It is finally time to start displaying the exceptionalism that is our birthright as humans, as the species we've baptized *Homo sapiens.* Unless and until we are able to do so, we will continue to put each other, ourselves, civil societies, and our planet at risk.

Acknowledgements

My birth certificate claims I was born in 1938 but the things I most value in myself didn't spring into being until the mid 1970s. Lionel Tiger and Robin Fox created a graduate department of anthropology on the Livingston campus of Rutgers University that encouraged a lively interdisciplinary exchange of knowledge. The seeds that became this book were planted there, nurtured by vigorous discussion with faculty and classmates. It all felt like a never-ending party. We laughed, talked, read, and debated—in the lecture halls and out—sometimes late into the night, and sometimes on ten-miles runs. A bit of an intellectual late bloomer, I was at least a decade older than my fellow grad students, but none of that mattered, nothing really mattered except an open-minded exploration of each other's ideas. And then, of course—as Sarah Zippin Shaffer, a friend from those years, recently reminded me—there were the *actual* parties. We did throw some truly great parties. My two young children and I rented groom's quarters in an old stable in horse country just north of New Brunswick, New Jersey. I fondly remember my two kids falling asleep at the kitchen table, unwilling to go to bed and leave the laughter and the burst of ideas that were both literally and figuratively swirling above their heads. Those were glorious years, the happiest time of my life. So, I begin my acknowledgements with sincere thanks to and a deep appreciation of both Lionel and Robin and all that their vision made possible.

Robert Trivers, the outside member of my dissertation committee, has had the greatest impact of all my mentors on my approach to evolutionary theory. He has influenced the way I think, and while he may not always agree with my conclusions, I will always be grateful to him for helping me see the world in a different light. Perhaps most important, it was a discussion with Bob Trivers that first made me aware of our very human talent for self-deception and deceit. That insight has become a central part of my understanding of human nature.

The danger in acknowledging all that others have done to make this book possible is that any such list is bound to be incomplete. The two who probably sacrificed the most in order that I might chase my dreams are my children, Ingrid Gallagher and Craig Fowler. There is no way to sufficiently thank them. Not ever. When Ingrid's daughter, Maeve Gallagher, was in third grade, she once asked me to describe grandmothers. I came up with the standard cozy image, a plump and cuddly cookie-baking source of hand-knit mittens. She looked at me in calm appraisal, shook her head, and said, "You're nothing like that." Even then I understood the deep love and pride that informed her statement. Craig's two children, Erin and Kyle Fowler, both now fully grown, share Maeve's ability to make me feel valued for who I am, not who or what I might be instead. Kyle and his wife, Mariah, have given me two gorgeous great-grandsons, David and Quentin. And then there are my sisters, Sandra Kirkpatrick and Gail Snyder, as well as a full range of relatives who stand there ready to cheer me on. My thanks to all of them.

A network of friends becomes its own kind of family. Everyone should have a friend like Sarah Shaffer, but she is in no way a reliable critic. It's not that she doesn't see my flaws; it's more that she cherishes them as part of my entertainment value. In addition to responding enthusiastically to every book I've ever written (some of them unpublished and truly awful), Sarah and I have shared joys and sorrows and secrets no one else knows. I couldn't have done this without her. Edie Gordon predates even Sarah and ticks off many of the same boxes. And then there is my Gang of Six: Susan Zelten, Amy Gardner, Kathleen Deasy, Carolyn Hirsch, and Nancy Stewart. Carol Feese, another friend from my Eagles Mere days, joins the list of those I don't see often but who have helped shape me. James Tyrone Lane understands creative passion and has been part of my life since he was six and I was forty-six. Catherine Prevost and I view the natural world through a shared lens. Jean Gould, Sharon Hamilton, and Vicki Coe are friends who are also writers and valued in both capacities.

I left Rutgers knowing that neither Charles Darwin nor I understood nearly enough about economics and spent the next twenty years learning about the discipline and engaging in economic activism. The friends I made during that period of my life have a fierce dedication to addressing social and economic injustice. The intensity of their commitment to a cause makes them less fun at parties than the Rutgers contingent but unbelievably supportive of my every effort. Alanna Hartzok has always pushed me to publish. Jeff Smith has read and commented on every page of this, my most recent work. Gib Halverson reminded me of the importance of Earth's bounty. Fred Harrison saw the significance of my focus on the link between human evolution and economics before I did and, in 2001, urged me to write an essay for a journal he founded. Years later, Cliff Cobb encouraged me to further develop those ideas in an article for *The American Journal of Economics and Sociology*. More recently, Cliff overruled my insistence that I was too old to undertake a book-length work of nonfiction and persuaded me to write *What's Sex Got To Do with It?* The result may not be quite what Cliff had in mind, but it would not have happened without him.

It wasn't until I moved to Arlington, Massachusetts in 2008 that my interest in evolutionary theory came roaring back. How could it not? Harvard, the

Radcliffe Institute, MIT, the Broad Institute, and the many free, public lectures those institutions make available were but a short ride away on the # 77 bus. Oh, the wealth of knowledge that was suddenly part of my world! Attending those presentations and exchanging ideas with other members of the audience and the speakers themselves (thank you, Jonathan Losos) was like finding an oasis when thirsty, a banquet when hungry.

Which brings me to the Harvard Book Store and the Boston Museum of Science's Science Book Club for the Curious. There have been weeks when I've spent more money on books than I did on groceries. Brain food. I've never begrudged those expenses and tend to think of them as membership dues in the world's two best social clubs. I cherish the friends I have made in both places. The names that follow are but a partial list. Daniel Rinehart, who facilitates the Science Book Club's discussions, inadvertently kept me writing when I had decided that I was simply an old fool who ought to know when to give up. At the end of one meeting, Daniel and I were leaving the building at the same time. He was heading for his bicycle and I was heading for the subway when I muttered something to the effect that I must be nuts to think I could write an update to Darwin's theory. Daniel looked at me, genuinely puzzled, and said, "But why not?" Thanks, Daniel. I headed home thinking, *Why not?* and got right back to work. Many members of the Science Book Club volunteered to read bits and pieces of the chapters when they were in process. Mika White not only read the whole thing, but she also edited it! I am sufficiently dyslexic that even a draft that I had proofread was an editor's nightmare. Mika did more than catch typos and errors in grammar and punctuation; she knew the science, spotted my misspelling of an obscure theorist's name, and asked terrific questions. "What does an evolutionary race to the finish look like?" Any errors that appear in this book happened after it left Mika's hands and were introduced by me in subsequent rewrites.

The Science Book Club exposed me to authors I might not have otherwise read and discussions of those ideas with other members of the group helped me discover missing pieces in the theory I was busy assembling. The partial list that follows reflects some of the regulars willing to read at least parts of my manuscript. Chris Geggis, Brian Abdow, Bic Stevens, Michele Sprengnether, Allie G. Durak, Reg McKeen, Karen Van Knapp, and Genevieve Berumen. I first met Genevieve at the Harvard Book Store's book club. It was she who suggested that I join the Science Book Club. I will always be grateful to her for that and for her friendship. Genevieve must hold some kind of record for the number of book clubs, currently seven, to which she simultaneously belongs. She also understands more about more things than anyone I have ever known. Meeting Genevieve for a bite to eat before attending the book group is a guaranteed celebration of friendship, food, and ideas.

I have always been fortunate in my friends. Deabru Gotson alerted me to the importance of the theories of Lynn Margulis and encouraged my own confidence and sense of fun. Joanne Blondin, a writer I met at the Harvard Book Store, first urged me to try to publish my thoughts on Darwin and then graciously looked over parts of the work and offered comments. Bill and Sue Bourne read random pages whenever a day's writing had entertained me enough that I couldn't wait to share an unanticipated insight and fired off emails to them. These two curious laypeople with no background in my field always asked illuminating questions. When I learned that they sometimes started their day debating a point I had made the night before, I allowed myself to believe that a general audience might find something of interest in my book. I met Paul Hackbarth walking back from Trader's Joe's on the Arlington bike path. I have no idea how it happened, but we almost immediately began discussing evolutionary theory. It was only when I realized that Paul had walked well beyond his usual turnoff and his ice cream was melting that we agreed to meet at Szechuan's Dumpling to continue the debate over a series of Monday lunches. I once pounded the table in such fury that I sheepishly returned to the restaurant the next day to apologize to the owners. I appreciate Paul for never expecting similar gestures of contrition. It was he who helped me understand that no one would ever listen if I started out by telling them Darwin was wrong. I wish I could say that I have succeeded in following Paul's advice. I've tried but, at best, have managed only to tone myself down. A bit.

And, finally, my thanks to Shepheard-Walwyn. The world is a better place because of independent bookstores and independent publishers. I'm grateful to Fred Harrison for introducing me to the folks at Shepheard-Walwyn, a publishing house whose stated mission is "to find books that have a lasting value, feeding heart and mind", and humbled to be considered worthy of inclusion in their list. Mirella Lombardo, Editorial Director at Shepheard-Walwyn, and Tracey Kerrigan, Marketing Manager, are a delight to work with, quickly responsive and professional in every way. I am deeply appreciative of everyone who has helped me bring this book to fruition. It goes without saying that any mistakes are fully mine. A special note of thanks goes to those who gave me the courage to risk making them.

Permissions

Endnotes

Preface

ix **Why hasn't the discovery of an astonishing evolutionary mutation—the end-to end fusion that combined great ape chromosomes. . . :** Jorge J. Yunis and Om Prakash, "The Origin of Man: A Chromosomal Fusion Pictorial Legacy," *Science* 215: 1525–1530.

ix **. . . the traits that were most significantly impacted by our speciation event were those relating to sex and the brain:** Manuel Ruiz Rejón, "The Origin of Human Species: A Chromosome Fusion," www.bbvaopenmind.com January 17, 2017.

x **In *The Sixth Extinction* Elizabeth Kolbert makes a compelling case for an extinction event of massive proportions:** Elizabeth Kolbert, *The Sixth Extinction* (New York: Picador, 2015).

xi **Darwin, himself, was one of the first to disavow the validity of the evidence those fossils offered:** Kolbert, Ibid., 79.

Chapter 1

1 **For the answer to that question we need to turn to Charles Darwin:** Charles Darwin, *The Descent of Man and Selection in Relation to Sex* (New York: Penguin Group, 2007).

2 **Prevailing theories of mate choice emphasized male to male competition:** Ferris Jabr, "How Beauty Is Making Scientists Rethink Evolution," *The New York Times Magazine* (January 9, 2019).

2 **Robert Trivers, the biologist whose research established the foundation:** Robert Trivers, "Sexual Selection and Resource Accruing Ability in Anolis Garmani," *Evolution 30(2)*: 256-269.

3 **Certain that women's sexual decisions were more complex:** Heather Remoff, "Female Choice: An Investigation into Human Breeding System Strategy," Ph.D. dissertation, Rutgers University, Ann Arbor Michigan Registration 692–164. University Microfilms International: 1980.

4 **This definition is not so different from the one Darwin gave:** Charles Darwin, *The Descent of Man and Selection in Relation to Sex* (New York; Penguin Group, 2007) 409.

7 **And since it was only in the early 1980s that scientists became aware of the fusion that combined great ape chromosomes 2A and 2b into human chromosome 2:** Ijdo, J. W.; Ward, D. C.; r, . Reeders, S. T.; and Wells, R.A., "Origin of Human Chromosome 2: an Ancestral Teleomere-Teleomere Fusion," (*Proceedings, National Academy of Sciences, USA* October 15: 88920): 9015-9055.

7 **"Rousseau repudiated Hobbs' animal analogy by stressing the significant difference in the reproductive physiology of human females.":** Quoted in Evelleen Richards, *Darwin and the Making of Sexual Selection* (Chicago: University of Chicago Press, 2017) 69.

8 **Every one of those books gave me some new angle to consider, but it was Terrence Deacon's:** Terrence W. Deacon, *The Symbolic Species: The Co-evolution of Language and the Brain* (New York: W. W. Norton &Company, 1997).

9 **"Therefore, accounting for the extravagant complexity of language in terms of sexual selection":** Terrence Deacon, "A Role for Relaxed Selection in the Evolution of Language Capacity," *PNAS* 107(2): 9000-9006.

Chapter 2

10 **Over 7,500 books have been written about Darwin:** David Dobbs, "Survival of the Prettiest." Book Review *New York Times* (September 18, 2017).

11 **Now raise the curtain on the Anthropocene**: Paul J. Crutzen, "Geology of Mankind," *Nature* 415 (2002): 23.

11 **What then are we to make of the rapid changes in beak and talon size recently documented:** Douglas Quenqua, "Things Looked Bleak Until These Birds Rapidly Evolved Big Beaks," *New York Times* (June 13, 2018).

12 **In other words, the very excess of a trait that researchers had documented as being highly attractive to females made it vulnerable to predation:** Michael J. Ryan, *A Taste for the Beautiful: The Evolution of Attraction* (Princeton: Princeton University Press, 2017) 64.

14 **It was a rare woman who had sex with a man who had not fed her first:** Heather Remoff, *Sexual Choice: A Woman's Decision* (New York: E. P. Dutton, 1984) 145.

15 **The Bajau people of Southeast Asia are such exceptionally skilled divers:** Carl Zimmer, " Bodies Remodeled for a Life at Sea," *New York Times* (April 24, 2018).

18 **In the opening pages of Michael Ryan's book, *A Taste for the Beautiful*, he describes observing the Resplendent Quetzal:** Op. cit., 9.

18 **Anyone who has heard lectures by Richard Prum:** Richard O. Prum, *The Evolution of Beauty: how Darwin's Forgotten Theory of Mate Choice Shapes the Animal World—and Us* (New York: Doubleday, 2017) 92.

22 **What appears to be going on with human females is that they find beauty where they have first seen other attributes:** Heather Remoff, Ibid., 116.

24 **Our path diverged from the Great Apes when two of their chromosomes fused into the one that is now ours:** Manuel Ruiz Rejón, "The Origin of Human Species: A Chromosome Fusion?" www.bbvaopenmind.com.

24 **One of the most fascinating examples of evolutionary speed is documented in research:** Jonathan B. Losos, *Improbable Destinies: Fate, Chance, and the Future of Evolution* (New York: Riverhead Books, 2017).

24 **"It turns out that Darwin and a century of biologists following him were wrong in one key aspect":** Losos, Ibid., 20.

24 **Early Research on sexual selection in anole lizards:** Robert Trivers, "Sexual Selection and Resource Accruing Ability in Anolis Garmani," *Evolution 30(2)*: 256-269.

Chapter 3

25 **I readily admit that I found my first exposure to the Hardy-Weinberg equation deeply unsettling.** C. C. Li, *Population Genetics* (Chicago: The University of Chicago Press, 1955) 3–11.

25 **Since Darwin drew so heavily on examples from animal husbandry:** Charles Darwin, *The Origin of Species* (New York: Penguin Books 1985).

26 **When Jonathan Losos first began investigating the adaptive properties:** Jonathan Losos, *Improbable Destinies: Fate, Chance, and the Future of Evolution* (New York: Riverhead Books, 2017).

26 **Genetic analysis fails to support a case for common ancestry between these look-alike lizards:** Ibid. 197.

27 **Since these changes had occurred in no more than a decade:** Ibid.,163.

28 **In only two generations, leg lengths were appreciably shorter:** Ibid., 169–179.

31 **Probably because of that, married couples are more genetically similar to each other:** www.theconversation.com/married-couples-have-similar-DNA

31 **She explained the immediate attraction she felt towards him by speculating it was due to mannerisms he shares:** Aaron Long, "First I Met My Children, Then My Girlfriend, They're Related," *New York Times* (September 28, 2018).

32 **Why don't children who have been reared together fall prey to the gene-recognition-pull of erotic desire:** Joseph Shepher, "Mate Selection Among Second Generation Kibbutz Adolescents and Adults: Incest Avoidance and Negative Imprinting," *Archives of Sexual Behavior.* 1(4): 293–307.

33 **Losos points out that since *E. coli* reproduce asexually, the six year timespan makes it pretty clear that sexual selection is not a necessary ingredient:** Jonathan Losos. Ibid., 228–234.

33 **My initial focus on the importance of resource-accruing ability as one of the factors human females evaluate:** Robert Trivers, "Sexual Selection and Resource Accruing Ability in Anolis Garmani," *Evolution 30(2)*: 256-269.

35 **Much as I might want my obituary-status-unknown professor to have a little talk with Mr. Darwin:** Rory Galloway, "Galapagos Finches Caught in the Act of Becoming a New Species," *BBC News* (www.bbc/news/science/environment 2017).

Chapter 4

37 **"It is indeed fortunate that the equal transmission of characters prevails":** Charles Darwin, *The Descent of Man and Selection in Relation to Sex* (New York: Penguin Group, 2007), 382.

37 **"In order that woman should reach the same standard as man":** Ibid., 383.

38 **Before deciding to marry his cousin:** Maria Popova, "Charles Darwin's List of the Pros and Cons of Marriage," (www.brainpickings.org/2012/darwin-list-pros-and-cons-of-marriage).

39 **Despite Mr. Darwin's belief that female choice was no longer operational in humans:** Heather Remoff, *Female Choice: A Woman's Decision* (New York: E. P. Dutton, 1984) 114.

40 **Alfred Russel Wallace informed Darwin that poorly adapted early humans:** Charles Darwin, *The Descent* Ibid., 200.

41 **As his preface to the second edition of *The Descent of Man:*** Charles Darwin, *The Decent of Man and Selection in Relation to Sex* (New York: A. L. Burt Company, 1874), iv.

42 **Epigenetic inheritance and the inheritance of acquired characteristics are not:** Jonathan Shaw, Is Epigenetics Inherited?" *Harvard Magazine* (2017), www.harvardmagazine.com/2017/05is-epigenetics-inherited.

43 **I was especially impressed when Karin Michels, Sc. D., a 2017 Radcliffe Fellow:** https://www.youtube.com/watch?v=gENf0DG9Xm0

Chapter 5

46 **The earliest known fossils with features that resemble anatomically modern humans:** Carl Zimmer, "How Did We Get to Be Human," *New York Times* (November 21, 2018).

46 **Christopher Walsh, a neurobiologist with expertise in genetic analysis:** Christopher A. Walsh, "Genes, Cognition, and Human Brain Evolution," Harvard Museum of Natural History Lecture (Cambridge, MA: October 10, 2018).

46 **Having long been drawn to Stephen Jay Gould's and Niles Eldredge's theory of punctuated equilibria:** Stephen Jay Gould, *The Panda's Thumb* (New York: W. W. Norton, 1980) 183. Niles Eldredge, *Time Frames: The Rethinking of Darwinian Evolution and the Theory of Punctuated Equilibria* (New York: Simon and Schuster, 1985).

46 **A study conducted in 1982 by Jorge Yunis and Om Prakash:** study described in Kenneth R. Miller, *The Human Instinct: How We Evolved to Have Reason, Consciousness, and Free Will* (New York: Simon and Schuster Paperbacks, 2019) 231 and Karen H. Miga, "Chromosome-Specific Centromere Sequences Provide an Estimate of the Ancestral Chromosome 2 Fusion Event in Human Genomes," *Journal of Heredity* 108 (1) 45–52 **and** Jorge J. Yunis and Om Prakash, "The Origin of Man: A Chromosomal Fusion Pictorial Legacy," *Science* 215: 1525–1530.

50 **"These genes are expressed, above all, in highly significant tissues and organs, such as the brain and the gonads."** Manuel Ruiz Rejón, "The Origin of Human Species: A Chromosome Fusion?" (www.bbvaopenmind.com. January 17, 2017).

51 **The sixty-six women I interviewed:** Heather Remoff, *Sexual Choice: A Woman's Decision* (New York: E. P. Dutton, 1984) 12 - 57.

51 **When a female chimp is in heat:** Melvin Konner, M. D., *Women After All: Sex, Evolution, and the End of Male Supremacy* (New York: Norton Paperback, 2016) 113.

Chapter 6

55 **Christopher Chabris and Daniel Simons, psychologists and co-authors of the book, *The Invisible Gorilla*:** Christopher Chabris and Daniel Simons, *The Invisible Gorilla and Other Ways Our Intuitions Deceive Us* (New York: Broadway Paperbacks, 2011). http://www.theinvisiblegorilla.com/gorilla_experiment.html.

56 **She not only convincingly described research that documented the ease with which false memories:** Elizabeth Loftus, "Investigating False Memories" (Cambridge: Radcliffe Institute Lecture, 2015). **and** Susan Seligson, "Investigating False Memories: Memory is Unreliable," (Cambridge: *Radcliffe Magazine,* May 2015).

58 **What possible benefit could be attached to such a defective understanding:** Lionel Tiger, *Optimism: The Biology of Hope* (New York: Simon & Schuster, 1979). **And** Robert Trivers, *The Folly of Fools: The Logic of Deceit and Self-Deception* (New York: Basic Books, 2011).

60 **A mark struck in stone had the authority:** Felix Martin. *Money: The Unauthorized Biography–from Coinage to Cryptocurrencies* (New York: Vintage Books, 2015) 26–30.

60 **There are those who maintain that the human brain is nothing more than a very sophisticated computer:** Daniel Dennett, *From Bacteria to Bach and Back: The Evolution of Minds* (New York: W. W. Norton & Company, 2017).

Chapter 7

62 **As a species, humans are obsessed with sex:** Adam Rutherford, *Humanimal: How Homo sapiens Became Nature's Most Paradoxical Creature* (New York: The Experiment, 2018) 101.

62 **. . . roughly a quarter of people's views on gender equality are more complicated:** Claire Cain Miller, "Americans Value Equality More at Work than at Home," *New York Times* (December 3, 2018).

63 **Early sociobiologists and current evolutionary psychologists give the nod to the cheap sperm/expensive egg:** Martin Daly and Margo Wilson, *Sex, Evolution & Behavior* (North Scituate, Massachusetts: Duxbury Press, 1978).

65 **So, one strategy that women employ is to "mate up":** Heather Remoff, *Sexual Choice: A Woman's Decision* (New York: E. P, Dutton, 1984).

67 **One of the big advantages of a system of language is that it makes the future conceptually available:** Heather Remoff , "Female Choice: An Investigation into Human Breeding System Strategies" (Ph. D. dissertation, 1980).

70 **Note the passive coyness in my phrase, "found ourselves":** Heather Remoff, *Sexual Choice.* Ibid., 21.

Chapter 8

72 **Manuel Ruiz Rejón, the Spanish researcher whose work with the fusion that could be either our original sin or our path to glory:** Manuel Ruiz Rejón, "The Origin of Human Species: A Chromosome Fusion?" (www.bbvaopenmind.com January 17, 2017).

72 **Lucy, officially designated *Australopithecus afarensis,* was not an ancestral human:** Matthew Warren, "Move Over, DNA: Ancient Proteins Are Starting to Reveal Humanity's History," *Nature. 256–264.* 2019.

73 **Our most recent shared ancestor with Chimpanzees was between 5 and 7 million years ago:** David Reich, *Who we are and How We got Here: ancient DNA and the New Science of the Human Past* (New York: Pantheon Books, 2018) 2.

74 **Language is what makes our unique brand of intelligence possible and any discussion of it must begin by making a clear distinction between language and communication:** Derek Bickerton, *Adam's Tongue: How Humans Made Language and How Language Made Humans* (New York: Hill and Wang, 2009).

77 **"More than 15 of the olfactory receptors that exist in the nose are also found in human skin cells":** Alex Stone, "Smell Turns Up in Unexpected Places," *New York Times* (October 13, 2019).

77 **No one has written a more compelling analysis of the evolution of language:** Terrence W. Deacon, *The Symbolic Species: The Co-evolution of Language and the Brain* (New York: W. W. Norton & Company, 1997).

78 **Deacon rejects what he calls hopeful monster theories:** Ibid., 35.

78 **"Biological change is vastly more slow and inflexible than language change":** Ibid., 110.

80 **To be human is to have a fully developed capacity for disingenuousness:** Robert Trivers, *The Folly of Fools: The Logic of Deceit and Self-Deception in Human Life* (New York: Basic Books, 2011)

81 **Recent research identifying olfactory receptors located throughout the human body:** Tatjana Abaffy, "Human Olfactory Receptors Expression and Their Role in Non-Olfactory Tissue–A Mini-Review," *Journal of Pharmacogenomics & Pharmacoproteomics* 6 (4): 10000152.

Chapter 9

84 **In 2018, a woman known only as Grace allowed her description of disastrous date with a popular stand-up:** https://babe.net/2018/01/13/aziz-ansari-28355

84 **"If you say you're a feminist, then fuck like a feminist":** Samantha Bee, https://www.hollywoodreporter.com/video/samantha-bee-addresses-aziz-ansari-accusations-thr-news-1075779

85 **In [a recent article] in *The Atlantic, Caitlin Flanagan says, "Sexual mores in the West have changed so rapidly…*":** Caitlin Flanagan, "The Humiliation of Aziz Ansari," *The Atlantic,"* January 14, 2016.

86 **When it comes to sex, women have a lot more power and control than they recognize:** Heather Remoff, *Sexual Choice* (New York: E. P. Dutton,1984) 8.

86 **The exercise of Female Choice flies mostly below our conscious awareness:** Ibid., 3–57.

88 **The phrases "soul mate" and true love" appear in his book so often:** Aziz Ansari and Eric Klinenberg, *Modern Romance; An Investigation* (New York: Penguin Books, 2015).

Chapter 10

92 **If Michael Ryan's premise about the seductive power of *more* is as true:** Michael Ryan, *A Taste for the Beautiful: The Evolution of Attraction* (Princeton: Princeton University Press, 2017) 48.

93 **In fairness to Darwin, he was worried that humans might be doing something wrong:** Charles Darwin, *The Descent of Man and Selection in Relation to Sex* (New York: Penguin Group, 2007) 214–215.

94 **The dramatic alterations in global ecosystems that define:** Elizabeth Kolbert, *The Sixth Extinction: An Unnatural History* (New York: Picador, 2105).

94 **Darwin's primary concern was not that humans were destroying the environment, but that the way:** Charles Darwin, Ibid., 217.

94 **"This is the doctrine of Malthus, applied to the whole animal and":** Charles Darwin, *The Origin of Species* (New York: Penguin Books, 1985) 68.

95 **Malthus himself never saw the need to document his treatise with actual measurements:** Robert Thomas Malthus, *An Essay on the Principle of Population as it Affects the Future of Society* (London: J. Johnson, 1798) 1.

96 **Darwin is obviously troubled:** Charles Darwin, *The Descent of Man,* (2007), Ibid., 214.

97 **"Mr. Wallace, in an admirable paper":** Ibid., 200.

98 **However, toward the end of Wallace's life, he rejected both the ideas of Malthus and the belief that natural selection:** Frederick Rockwell, "The Last of the Great Victorians; Special Interview with Dr. Alfred Russel Wallace," *Millgate Monthly* 7: 657–663. https://people.wku./charles.smith/wallace.

98 **That said, although Wallace continued to distance himself from Darwin's theory of sexual selection:** Evelleen Richards, *Darwin and the Making of Sexual Selection* (Chicago: University o f Chicago Press, 2017) 510.

100 **Wallace was sufficiently impressed by the theories of Henry George:** Alfred Russel Wallace, "Letter to Charles Darwin." July 9, 1881. https://www.darwinproject.ac.uk./letter.

100 **George used the term "land" to refer to *all* naturally occurring resources:** Henry George, *Progress and Poverty* (New York: Robert Schalkenbach Foundation, 1979) 38.

100 **Mr. Greg correctly anticipated the future disparity in land ownership:** Charles Darwin, *The Descent* (New York: Penguin Group, 2007) 215.

Chapter 11

103 **They believed the Irish were simply having too many children:** Darwin, Ibid., 215.

105 **Enter money, a glorious innovation that keeps track of who owes what:** Felix Martin, *Money - the Unauthorized Biography from Coinage to Cryptocurrencies* (New York: Vintage Trade Paperbacks, 2015).

106 **Robert Trivers differentiated between the resource-accruing ability:** Robert Trivers, "Sexual Selection and Resource Accruing Ability in Anolis Garmani," Ibid., 257.

106 **. . . we need to take into account the distinction between resource-accruing *bility* (RAA) and resource-accruing *opportunity* (RAO):** Heather Remoff, "Malthus, Darwin, and The Descent of Economics," *American Journal of Economics and Sociology* 75(4): 888.

109 **It was the failure of the fossil record to demonstrate the gradual change envisioned by Darwin:** Stephen Jay Gould, "The Episodic Nature of Evolutionary Change," *The Panda's Thumb* (New York: W. W. Norton & Company, 1980) 179–185.

109 **Eldredge's book:** Niles Eldredge, *Time Frames: The Rethinking of Darwinian Evolution and the Theory of Punctuated Equilibria* (New York: Simon and Schuster, 1985)121.

111 **The jawbone of the truly enormous baleen whale:** Nick Pyenson, *Spying on Whales: The Past, Present and Future of Earth's Most Amazing Creatures* (New York: Penguin Random House LLC, 2018).

Chapter 12

114 **Digital dating platforms make it possible to gather a wealth of data:** Maya Salam, "For Online Daters, Women Peak at 18 While Men Peak at 50, Study Finds. Oy," *New York Times* (August 15, 2018).

115 **The most popular *New York Times* article:** Alain de Botton, "Why You'll Marry the Wrong Person," *New York Times, Sunday Review* (May 28, 2016).

120 **In a demonstration of the mathematical calculations behind such tabulations of inclusive fitness:** Richard McElreath and Robert Boyd, *Mathematical Models of Social Evolution: A Guide for the Perplexed* (Chicago: University of Chicago Press, 2007) 82.

Chapter 13

126 **I've seen photographs of Australian termite "castles" that are close replicas:** Daniel C. Dennett, *From Bacteria to Bach and Back: The Evolution of Minds* (New York: W. W. Norton & Company, 2017) 238.

129 **I find it impossible to view the notion of race as anything but another of those artificial human constructs:** Stephen Pinker, *The Language Instinct: How the Mind Creates Language* (New York: Harper Perennial Modern Classics, 2007) 448.

Chapter 14

132 **I wish we'd listened to Al Gore in 2006:** Al Gore, "The Climate Crisis Is the Battle of Our Time, and We Can Win," *New York Times* (September 20, 2019).

135 **George's theory is even more relevant in the age of climate change than it was when he:** Henry George, *Progress and Poverty* (New York: Robert Schalkenbach Foundation, 1979).

135 **If anyone continues to doubt the kinds of economic distortions and injustices:** Eric Lipton and Jesse Drucker, "Symbol of 80s Greed Stands to Profit From Trump Tax Breaks for Poor Areas," *New York Times* (October 16, 2019).

138 **The dollar amount of the economic rents flowing from natural resources is enormous:** Scott Baker, "Land Worth = GDP and Enough to Replace All Taxes with Land Rent," https://www.opednews.com/articles/Land-Worth-GDP-and-enough-by-Scott-Baker-Land-Rights_Land-Value-Tax_Land-Value-Taxation_Taxation-160321-463.html **And** Mason Gaffney, "The Unplumbed Revenue Potential of Land," *Mason Gaffney Reader: Essays on Solving the Unsolvable* (New York: Henry George Institute, 2013) 83–112. **And** Fred Harrison, *Free Riding: Rent, the Social Galaxy and the Crime Against Humanity* (London: forthcoming, 2020. **And** Jeffrey Johnson Smith, *Counting Bounty: The Quest to Know the Worth of Land in America* (Waterville, Oregon: Trine Day publishing, 2020).

139 **The Windigo of Ojibwe legend is a monster made fearsome by insatiable desires:** Robin Wall Kimmerer, *Braiding Sweetgrass: Indigenous Wisdom, Scientific Knowledge, and the Teachings of Plants* (Minneapolis: Milkweed Editions, 2013) 305.

Chapter 15

142 **"In an essay describing hunter-gatherer peoples with few possessions as the original affluent society":** Marshall Sahlins as quoted by Robin Wall Kimmerer, *Braiding Sweetgrass* (Minneapolis: Milkweed Editions, 2013) 376.

142 **Each of us comes from a people who were once indigenous**: Ibid., 377.

142 **Ingrid joined me in attendance at a Radcliffe lecture on tool use in New Caledonian Crows:** Christian Rutz lecture covered by Ryan Mulcahy, "Some Crows Are Birds, Others Are Artists," *The Magazine of the Radcliffe Institute for Advanced Study* (Winter 2020) 9–11.

Chapter 16

148 **King James I formed the Virginia Company of London in1606, but forging a life in the new world:** Resource Library Entry, "The Plantation System" *National Geographic Society* (June 20, 2019).

149 **Let's start by comparing the amount of wealth held by white families to that held by Black:** Akilah Johnson, "That Was No Typo: The Median Net Worth of Black Bostonians Really Is $8," *Boston Globe* (December 11, 2017).

149 **"The ownership disparity is. . .** ***responsible*** **for a big share of the racial wealth gap":** Noah Smith, "Land Is Underrated as a Source of Wealth," *Bloomberg* (January 2, 2018).

150 **General Sherman's Special Field Order 15 proclaimed that enormous parcels:** Nikole Hannah-Jones, "What Is Owed: Without Economic Justice, There Can Be No True Equality," (*New York Times,* June 30, 2020).

150 **Not only did the freed people fail to retain those forty acre plots:** Nikole Hannah-Jones, Editor. *The 1619 Project* (*New York Times Magazine*, August 14, 2019).

151 **The more money a district spends on its schools, the better the schools:** Sergio Paçanha, "These Numbers Show That Black and White People Live in Two different Americas," *The Washington Post* (June 26, 2020).

151 **When it comes to funding public services, it's the owners of corporate, commercial, and investment properties who have been getting away with murder:** Christopher Ingraham, "American Land Barons," *Washington Post* (December 21, 2017).

151 **. . . nearly 30 million acres of U. S. farmland are held by foreign investors:** Renee Wilde, "American Soil Is Increasingly Foreign Owned," *NPR Report, All Things Considered* (May 27, 2019). **And** Jennifer Rudden, "US Foreign Investment," *Statista* (February 17, 2020). **And** Matt Levin, "Data Dig: Are Foreign Investors Driving Up Real Estate in Your California Neighborhood," *Cal Matters* (March 7, 2018).

.

Selected Bibliography

Abaffy, Tatjana. (2015). "Human Olfactory Receptors Expression and Their Role in Non-Olfactory Tissues–A Mini-Review." *Journal of Pharmacogenomics & Pharmacoproteomics.* 6(4): 10000152.

Adams, Martin. (2015) *Land: A New Paradigm for a Thriving World.* Berkeley, CA: North Atlantic Books.

Ansari, Aziz and Eric Klinenberg. (2015) *Modern Romance: An Investigation.* New York: Penguin Books.

Baker, Scott. (2016). "Land Worth = GDP and Enough To Replace All Taxes with Land Rent." https://www.opednews.com/articles/Land-Worth-GDP-and-enough-by-Scott-Baker-Land-Rights_Land-Value-Tax_Land-Value-Taxation_Taxation-160321-463.html

Berwick, Robert C., and Noam Chomsky. (2016). *Why Only Us: Language and Evolution.* Cambridge, Massachusetts: MIT Press.

Bickerton, Derek. (2009). *Adam's Tongue: How Humans Made Language and How Language Made Humans.* New York: Hill and Wang.

———. (2014). *More Than Nature Needs: Language, Mind, and Evolution.* Cambridge, Massachusetts: Harvard University Press.

Boston Globe. (December 11, 2017). "That Was No Typo: The Median Net Worth of Black Bostonians Really Is .$8." By Akilah Johnson.

Bregman, Rutger. ([2104] 2018). *Utopia For Realists: How We Can Build the Ideal World.* New York: Back Bay Books/ Little, Brown and Company.

———. (2020). *Humankind: A Hopeful History.* New York. Little, Brown and Company.

Cal Matters. (March 7, 2018). "Data Dig: Are Foreign Investors Driving Up Real Estate in Your California Neighborhood?" By Matt Levin.

Campbell, Bernard, Ed. (1972). *Sexual Selection And The Descent Of Man 1871–1971.* Chicago: Aldine Publishing Company.

Chabris, Christopher and Daniel Simons. ([2010] 2011). "The Invisible Gorilla and Other Ways Our Intuitions Deceive Us." New York: Broadway Paperbacks. http://www.theinvisiblegorilla.com/gorilla_experiment.html.

Crutzen, Paul J. (2002). "Geology of Mankind." *Nature* 415: 23

Daly, Martin and Margo Wilson. (1978). *Sex, Evolution & Behavior.* North Situate, Massachusetts: Duxbury Press.

Darwin, Charles. ([1845] 1906). *The Voyage of the H.M.S. Beagle round the World.* London: J.M. Dent & Sons Ltd.

———. ([1859] 1860). *Origin of Species.* New York: Hurst & Company.

———. ([1859] 1985). *Origin of Species.* New York: Penguin Books.

———. ([1871]1874). *The Descent of Man and Selection in Relation to Sex.* New York: A.L. Burt Company.

———. ([1871] 2007). *The Descent of Man and Selection in Relation to Sex.* New York: Penguin Group.

———. ([1876]1887). *The Life and Letters of Charles Darwin.* London: Charles Murray.

——. (1881). "Letter to Alfred Russel Wallace." July 12, https://www.darwinprojsect.ac.uk/letter.

——. (1996). *Charles Darwin's Letters: A Selection*. Ed. by Frederick Burkhardt. New York: Cambridge University Press.

Dawkins, Richard. (1976). *The Selfish Gene*. New York: Oxford University Press.

Deacon, Terrence W. (1997). *The Symbolic Species: The Co-evolution of Language and the Brain*. New York: W.W. Norton & Company.

——. (2010). "A Role for Relaxed Selection in the Evolution of the Language Capacity." *PNAS*. 107 (2): 9000-9006.

Dennett, Daniel C. (2017). *From Bacteria to Bach and Back: The Evolution of Minds* New York: W.W. Norton & Company.

de Waal, Frans. (2019). *Mamma's Last Hug: Animal Emotions and What They Tell Us about Ourselves*. New York: W.W. Norton & Company.

Diamond, Jared. ([1992] 2006). *The Third Chimpanzee: The Evolution and Future of the Human Animal*. New York: Harper Perennial.

Dobbs, David. (2017). "Survival of the Prettiest." Review of *Purpose and Desire*, by J. Scott Turner, *Darwin's Backyard* by James Costa, *Collecting Evolution* by Matthew J. James, *Darwin's First Theory* by Rob Wesson, and *The Evolution of Beauty* by Richard Prum. *New York Times*. September 18.

Eldredge, Niles. (1985). *Time Frames: The Rethinking of Darwinian Evolution and the Theory of Punctuated Equilibria*. New York: Simon and Schuster.

——. (1999). *The Pattern of Evolution*. New York: W. H. Freeman and Company.

Everett, Daniel L. (2009). *Don't Sleep, There Are Snakes: Life and Language in the Amazonian Jungle*. New York: Vintage Books.

——. (2012). *Language: The Cultural Tool*. New York: Vintage Books.

——. (2116). *Dark Matter of the Mind. The Culturally Unarticulated Unconscious*. Chicago: The University of Chicago Press.

——. (2017). *How Language Began: The Story of Humanity's Greatest Invention. New York: W. W. Norton & Company.*

Flanagan, Caitlin. (2016). "The Humiliation of Aziz Ansari." *The Atlantic*. January 14.

Fox, Robin and Lionel Tiger. ([1971] 1974). *The Imperial Animal*. New York: Dell Publishing Company, Inc.

Gaffney, Mason. (2013). "The Unplumbed Revenue Potential of Land." In *The Mason Gaffney Reader: Essays on Solving the "Unsolvable"* by Mason Gaffney, 83-112. New York: Henry George Institute.

Galloway, Rory. (2017). "Galapagos Finches Caught in the Act of Becoming New Species." *BBC News*: https://www.bbc/news/science/environment.

George, Henry. ([1879] 1979). *Progress and Poverty*. New York: Robert Schalkenbach Foundation.

——. ([1886] 1980). *Protection or Free Trade*. New York: Robert Schalkenbach Foundation.

Godfrey-Smith, Peter. (2016). *Other Minds: The Octopus, the Sea and The Deep Origins of Consciousness*. New York: Farrar, Straus, and Giroux.

Gould, Stephen Jay. (1980). *The Panda's Thumb: More Reflections in Natural History*. New York: W.W. Norton & Company.

——. (1981). *The Mismeasure of Man*. New York: W.W. Norton & Company.

——. (1991). *Bully for Brontosaurus: Reflections in Natural History*. New York: W. W. Norton & Company.

Greer, Germaine. (1985). *Sex and Destiny: the Politics of Human Fertility*. London: Picador.

Hannah-Jones, Nikole. Editor. (2019). *The 1619 Project* (*New York Times*, August 14).

———. (2020). "What Is Owed: Without Economic Justice, There Can Be No True Equality," (*New York Times*, June 30, 2020).

Harari, Yuval Noah. (2018). *Sapiens: A Brief History of Humankind.* New York: Harper Perennial.

Harrison, Fred. (1983). *The Power in the Land.* London: Shepheard-Walywn.

———. (2015), *As Evil Does.* London: Shepheard-Walwyn.

———. (2020). *Free Riding: Rent, the Social Galaxy and the Crime Against Humanity.* (forthcoming).

Ijdo, J. W.; A. Baldini; D. C. Ward; S. T. Reeders; and R. A. Wells. (1991). "Origin of Human Chromosome 2: an Ancestral Teleomere-Teleomere Fusion." *Proceedings, National Academy of Sciences, USA:* Oct 15; 88(20): 9015-9055.

Institute for New Economic Thinking. (June 10, 2020) "If You Want justice for Black Americans, You have to Fix This." Lynn Parramore interviews Derrick Hamilton.

Jabr, Ferris. (January 9, 2019). "How Beauty Is Making Scientists Rethink Evolution." *The New York Times Magazine.*

Kenneally, Christine. (2007). *The First Word: The Search for the Origins of Language.* New York: Penguin Books.

Kimmerer, Robin Wall. (2013) *Braiding Sweetgrass: Indigenous Wisdom, Scientific Knowledge, and the Teachings of Plants.* Minneapolis: Milkweed Editions.

Kolbert, Elizabeth. ([2014] 2015) *The Sixth Extinction: An Unnatural History.* New York: Picador.

Konner, Melvin, M. D. (2016) *Women After All: Sex, Evolution, and the End of Male Supremacy.* New York: W. W. Norton & Company.

Li, C. C. (1955) *Population Genetics.* Chicago: The University of Chicago Press.

Loftus, Elizabeth. (2015) "Investigating False Memories." Lecture presented at the Radcliffe Institute For Advanced Study, Cambridge, MA, May 14.

Losos, Jonathan B. (2017). *Improbable Destinies: Fate, Chance, and the Future of Evolution.* New York: Riverhead Books.

McElreath, Richard and Robert Boyd. (2007). *Mathematical Models of Social Evolution: A Guide for the Perplexed.* Chicago: University of Chicago Press.

Malthus, Thomas Robert. (1798). *An Essay on the Principle of Population, as it Affects the Future Improvement of Society.* London: J. Johnson.

Martin, Felix. (2015). *Money: The Unauthorized Biography–From Coinage to Cryptocurrencies.* New York: Vintage Trade Paperbacks.

Maynard Smith, John. (1971). "What Use Is Sex." *Journal of Theoretical Biology.* 30 319-335.

Michels, Karen. (2017) "Is Epigenetics Inherited?" Radcliffe Fellows Presentation, Cambridge, MA. https://www.youtube.com/watch?v=gENf0DG9Xm0

Miga, K. H.. (2016). "Chromosome-Specific Centromere Sequences Provide an Estimate of the Ancestral Chromosome 2 Fusion Event in Hominin Genomes." *Journal of Heredity.* 108 (1) 45-52.

Miller, Kenneth R. (2019). *The Human Instinct: How We Evolved to Have Reason, Consciousness, and Free Will.* New York: Simon and Schuster Paperbacks.

Mulcahy, Ryan. "Some Crows are Birds, Others Are Artists." (On Christian Rutz and the New Caledonian Crow.) *Radcliffe: The Magazine of the Radcliffe Institute of Advanced Study,* Winter 2020, pp. 9–11.

National Public Radio. All Things Considered. (May 27, 2019). "American Soil Is Increasingly Foreign Owned." By Renee Wilde.

National Geographic Society. (June 20, 2019). "The Plantation System." www.nationalgeographic.org/encyclopedia/plantation-system.

New York Times.
———. (October 13, 2014). "Smell Turns Up in Unexpected Places." By Alex Stone.
———. (May 28, 2016). "Why You'll Marry the Wrong Person." By Alain de Botton.
———. (April 24, 2018). "Bodies Remodeled for a Life at Sea." By Carl Zimmer.
———. (June 13, 2018). "Things Looked Bleak Until These Birds Rapidy Evolved Big Beaks." By Douglas Quenqua.
———. (August 15, 2018). "For Online Daters, Women Peak at 18 While Men Peak at 50, Study Finds. Oy." By Maya Salam.
———. (September 28, 2018). "First I Met My Children, Then My Girlfriend. They're Related." By Aaron Long.
———. (November 8, 2018). "In Cave in Borneo Jungle, Scientists Find Oldest Figurative Painting in the World". By Carl Zimmer.
———. (November 21, 2018) "How Did We Get to Be Human?" By Carl Zimmer.
———. (December 3, 2018). "Americans Value Equality at Work More Than Equality at Home" By Claire Cain Miller.
———. (December 14, 2018). "Neanderthals, the World's First Misunderstood Artists." By Carl Zimmer.
———. (January 30, 2019). "High Ceilings and a Lovely View: Denisova Cave Was Home to a Lost Branch of Humanity." By Carl Zimmer.
———. (September 20, 2019). "The Climate Crisis is the Battle of Our Time, and We Can Win." By Al Gore.
———. (October 26, 2019). "Symbol of '80s Greed Stands to Profit From Trump Tax Break for Poor Areas." By Eric Lipton and Jesse Drucker.
———. (June 24, 2020). "What Is Owed: Without Economic Justice There Can Be No TrueEquality."www.nytimes.com/interactive/2020/06/magazine/reparations-slavery.html
Ostler, Jeffrey. "The Shameful Final Grievance of the Declaration of Independence." *The Atlantic,* February 8, 2020.
Pyenson, Nick. (2018). *Spying on Whales: The Past, Present ,and Future of Earth's Most Awesome Creatures.* New York: Penguin Random House LLC.
Pinker, Steven. (2007). *The Language Instinct: How the Mind Creates Language.* New York: Harper Perennial Modern Classics.
Popova, Maria. (1938 [2012]). "Charles Darwin's List of the Pros and Cons of Marriage."https://www.brainpickings.org/2012/08/14/darwin-list-pros-and-cons-of-marriage.
Prum, Richard O. (2017). *The Evolution of Beauty: How Darwin's Forgotten Theory of Mate Choice Shapes the Animal World—and Us.* New York: Doubleday.
Quammen, David. (2018). *The Tangled Tree: A Radical New History of Life.* New York: Simon & Schuster Paperbacks.
Raup, David M. (1991) *Extinction: Bad Genes or Bad Luck?* New York: W. W. Norton & Company.
Reich, David. (2018). *Who We Are and How We Got Here: Ancient DNA and the New Science of the Human Past.* New York: Pantheon Books.
Rejón. Manuel Ruiz. (2017). "*The Origin of Human Species: a Chromosome Fusion?*" www.bbvaopenmind.com.
Remoff, Heather Trexler. (1980). "Female Choice: An Investigation into Human Breeding System Strategy." Ph.D. dissertation. Rutgers University. Registration 692-164. Ann Arbor, Michigan: University Microfilms International.
———. (1984). *Sexual Choice: A Woman's Decision: Why and How Women Choose the Men They Do as Sexual Partners.* New York: E. P. Dutton.

———. (2001). "Sex, Economics, and Evolution:" In *Geophilos: Sex, Economics, and the Descent of Man,* edited by Fred Harrison. UK: Land Research Trust.

———. (2016). "Malthus, Darwin, and the Descent of Economics." *American Journal of Economics and Sociology.* 75(4): 862-903.

Richards, Evelleen. (2017). *Darwin and the Making of Sexual Selection.* Chicago: University of Chicago Press.

Rockwell, Frederick. (1912). "The Last of the Great Victorians: Special Interview with Dr. Alfred Russel Wallace." *Millgate Monthly* 7: 657-663. https://people.wku./charles.smith/wallace.

Rutherford, Adam. (2019). *Humanimal: How Homo sapiens Became Nature's Most Paradoxical Creature. A New Evolutionary History.* New York: The Experiment, LLC.

Ryan, Michael J. (2017). *A Taste for the Beautiful: The Evolution of Attraction.* Princeton: Princeton University Press.

Sapolsky, Robert M. (2017). *Behave: The Biology of Humans at Our Best and Worst.* New York: Penguin Press.

Seligson, Susan. (2015). "Investigating False Memories: Memory is Unreliable." *Radcliffe Magazine.* May 14.

Shaw, Jonathan. (2017). "Is Epigenetics Inherited?" *Harvard Magazine.* https:www.harvardmagazine.com/2017/05is-epigenetics-inherited.

Shepher, Joseph. (1971). "Mate Selection Among Second Generation Kibbutz Adolescents and Adults: Incest Avoidance and Negative Imprinting." *Archives of Sexual Behavior.* 1(4): 293-307.

Smith, Charles H. (1999). "Alfred Russel Wallace on Evolution: A Change of Mind?" Presented February 26 at the Symposium on the History of Medicine and Science at the University of Of Southern Mississippi, Hattiesburg, MS. https://people.wku.edu/Charles.smith/essays/WALLTALK.htm.

Smith, Jeffrey Johnson Smith. (2020). *Counting Bounty: The Quest to Know the Worth of Land in America.* Walterville, Oregon: Trine Day Publishing.

Smith, Noah. "Land Is Underrated as a Source of Wealth." *Bloomberg.* January 2, 2018.

Smithsonian Annual III. (1917). *Man and Beast: Comparative Social Behavior.* City of Washington: Smithsonian Institution Press.

Statista. (February 17, 2020). "US Foreign Investment." By Jennifer Rudden.

Tiger, Lionel and Joseph Shepher. (1975). *Women in the Kibbutz.* New York: Harcourt Brace Jovanovich.

Tiger, Lionel. (1979). *Optimism: The Biology of Hope.* New York: Simon & Schuster.

Trivers, Robert L. (1972). "Parental Investment and Sexual Selection." In *Sexual Selection and the Descent of Man, 1871-1971,* edited by Bernard Campbell, 136-179. Chicago: Aldine Publishing Company.

———. (1976). "Sexual Selection and Resource Accruing Ability in Anolis Garmani." *Evolution.* 30(2): 256-269.

———. (2011). *The Folly of Fools: The Logic of Deceit and Self-Deception in Human Life.* New York: Basic Books.

Wallace, Alfred Russel. (1880). "How to Nationalize the Land: A Radical Solution of the Irish Problem." *Contemporary Review* 38: 716-736.

———. (1881). "Letter to Charles Darwin." July, 9. https://www.darwinproject.ac.uk./letter.

———. (1883). "The Why and How of Land Nationalization." *Macmillans Magazine* 48: 357-368,485-493. http://people.wku.edu/charlessmith/wallace/S365.htm.

Wallace, Robert A. (1979). *Animal Behavior: Its Development, Ecology, and Evolution.* Santa Monica, California: Goodyear Publishing Company, Inc.

Wallace-Wells, David. (2019). *The Uninhabitable Earth: Life After Warming.* New York: Tim Duggan Books.

Walsh, Christopher A., M. D. (2018). "Genes, Cognition and Human Behavior." Public Lecture at Harvard Museum of Natural History, Cambridge, MA, Oct.10.

Warren, Matthew. (2019). "Move Over, DNA: Ancient Proteins Are Starting to Reveal Humanity's History." *Nature. 256–264.* www.nature.com.newsfeature.

Washington Post. (December 21, 2017) "American Land Barons." By Christopher Ingraham.

———. (June 23, 2020) "These Numbers Show That Black and White People Live in Two Different Americas." By Sergio Paçanha.

Williams, George C. (1966). *Adaptation and Natural Selection.* Princeton: Princeton University Press.

———. (1975). *Sex and Evolution.* Princeton: Princeton University Press.

Wilson, A. N. (2017). *Charles Darwin: Victorian Mythmaker.* New York: Harper Collins.

Wilson, David Sloan. (2019). *This View of Life: Completing the Darwinian Revolution.* New York: Pantheon Books.

Wolfe, Tom. (2016). *The Kingdom of Speech.* New York: Little Brown and Company.

Wulf, Andrea. (2016). *The Invention of Nature: Alexander von Humboldt's New World.* New York: Vintage Books.

Yunis, Jorge J., and Om Prakash. (1982). "The Origin of Man: A Chromosomal Pictorial Legacy." *Science* 215: 1525-1530.

Index